A2 UNIT 2

STUDENT GUIDE

CCEA

Biology

Biochemistry, genetics and evolutionary trends

John Campton

Hodder Education, an Hachette UK company, Blenheim Court, George Street, Banbury, Oxfordshire OX16 5BH

Orders

Bookpoint Ltd, 130 Park Drive, Milton Park, Abingdon, Oxfordshire OX14 4SE

tel: 01235 827827

fax: 01235 400401

e-mail: education@bookpoint.co.uk

Lines are open 9.00 a.m.–5.00 p.m., Monday to Saturday, with a 24-hour message answering service. You can also order through the Hodder Education website: www.hoddereducation.co.uk

ISBN 978-1-4718-6399-8

First printed 2017

Impression number 5 4 3 2 1

Year 2021 2020 2019 2018 2017

Cover photo: andamanse/Fotolia; other photos: p. 82, Dr George Chapman, Visuals Unlimited/ SPL

Typeset by Integra Software Services Pvt. Ltd, Pondicherry, India

Printed in Slovenia

Contents

Content Guidance

Questions & Answers

Getting the most from this book

Exam tips

Advice on key points in the text to help you learn and recall content, avoid pitfalls, and polish your exam technique in order to boost your grade.

Knowledge check

Rapid-fire questions throughout the Content Guidance section to check your understanding.

Knowledge check answers

1 Turn to the back of the book for the Knowledge check answers.

Summaries

- Each core topic is rounded off by a bullet-list summary for quick-check reference of what you need to know.

Exam-style questions

Commentary on the questions

Tips on what you need to do to gain full marks, indicated by the icon ⓔ

Sample student answers

Practise the questions, then look at the student answers that follow.

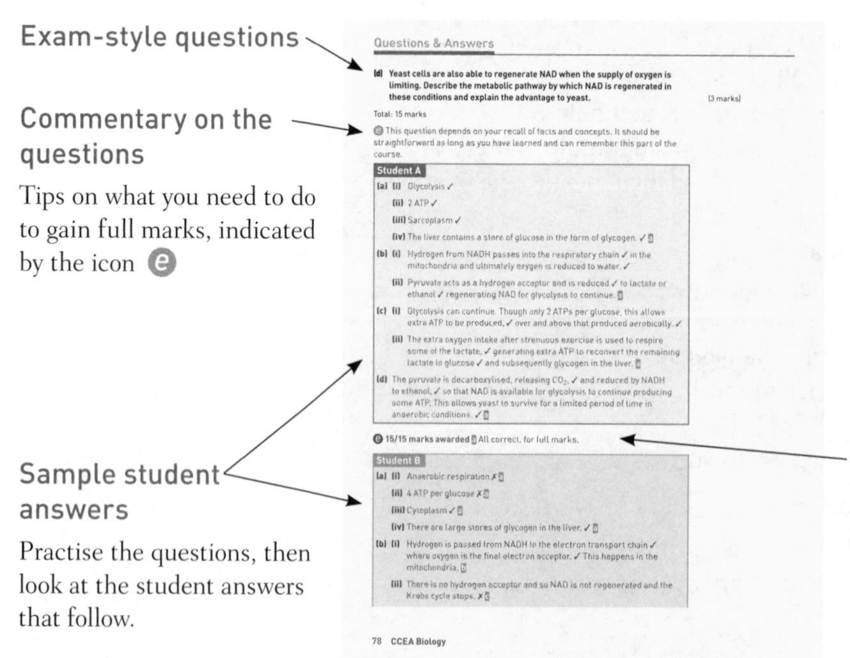

Questions & Answers

(d) Yeast cells are also able to regenerate NAD when the supply of oxygen is limiting. Describe the metabolic pathway by which NAD is regenerated in these conditions and explain the advantage to yeast. (3 marks)

Total: 15 marks

ⓔ This question depends on your recall of facts and concepts. It should be straightforward as long as you have learned and can remember this part of the course.

Student A

(a) (i) Glycolysis ✓

(ii) 2 ATP ✓

(iii) Sarcoplasm ✓

(iv) The liver contains a store of glucose in the form of glycogen. ✓

(b) (i) Hydrogen from NADH passes into the respiratory chain ✓ in the mitochondria and ultimately oxygen is reduced to water. ✓

(ii) Pyruvate acts as a hydrogen acceptor and is reduced ✓ to lactate or ethanol ✓ regenerating NAD for glycolysis to continue.

(c) (i) Glycolysis can continue. Though only 2 ATPs per glucose, this allows extra ATP to be produced, ✓ over and above that produced aerobically. ✓

(ii) The extra oxygen intake after strenuous exercise is used to respire some of the lactate, ✓ generating extra ATP to reconvert the remaining lactate to glucose ✓ and subsequently glycogen in the liver.

(d) The pyruvate is decarboxylised, releasing CO_2, ✓ and reduced by NADH to ethanol, ✓ so that NAD is available for glycolysis to continue producing some ATP. This allows yeast to survive for a limited period of time in anaerobic conditions. ✓

ⓔ **15/15 marks awarded** All correct, for full marks.

Student B

(a) (i) Anaerobic respiration ✗

(ii) 4 ATP per glucose ✗

(iii) Cytoplasm ✓

(iv) There are large stores of glycogen in the liver. ✓

(b) (i) Hydrogen is passed from NADH to the electron transport chain ✓ where oxygen is the final electron acceptor. ✓ This happens in the mitochondria.

(ii) There is no hydrogen acceptor and so NAD is not regenerated and the Krebs cycle stops. ✗

78 CCEA Biology

Commentary on sample student answers

Read the comments (preceded by the icon ⓔ) showing how many marks each answer would be awarded in the exam and exactly where marks are gained or lost.

About this book

This book will help you to prepare for the A2 Unit 2 examination for CCEA Biology.

The **Content Guidance** contains everything that you should learn to cover the specification content of A2 Unit 2. It should be used as a study aid as you meet each topic, when you prepare for end-of-topic tests, and during your final revision. For each topic there are *exam tips* and *knowledge checks* in the margins. Answers to the knowledge checks are provided towards the end of the book. At the end of each topic there is a list of the *practical work* with which you are expected to be familiar. This is followed by a comprehensive, yet succinct, *summary* of the points covered in each topic.

The **Questions & Answers** section contains questions on each topic. There are answers written by two students together with comments on their performances and how they might have been improved. There is a range of question styles of the kind that you will encounter in the A2 Unit 2 exam, and the students' answers and comments should help with your examination technique.

Developing your understanding

The key to effective study is to make it *active*. For example, you can take the information given in this book and present it in different ways, such as bullet-point lists to summarise the key points, flash cards, annotated diagrams to show structure and function, annotated graphs, and spider diagrams or mind maps. The important thing is that you are *doing* something and so thinking about the topic in detail. This increases the chance of you remembering it and also allows you to see links between different areas of biology. This kind of deep understanding is the key to getting top marks in the exam.

Writing essays on different topics will consolidate your understanding and will give you practice at Section B questions.

Biology has a huge number of specialist terms and it is important that you use them accurately. Improve your understanding by compiling your own glossary of terms for each topic. Key terms are shown in **bold** (with a few in the margin). For each you should provide a definition.

Think about the information in this guide, so you are able to *apply* your understanding in unfamiliar situations. It is useful to read around the subject using a variety of resources.

Synoptic links

In order to develop your understanding of the subject as a whole, you need to work at making connections between the topics you have studied so far. It is essential that you revisit the core concepts that you learned at AS since these often underlie the topics at A2. For example, many areas of biology rely on an understanding of cell biology (ultrastructure and function) since cells are the 'units of life'. Other synoptic links are highlighted in the Content Guidance.

The specification, past papers and other useful documents can be accessed at www.ccea.org.uk.

Content Guidance

Respiration

Energy, adenosine triphosphate and respiration

Energy is defined as 'the capacity to do work'. The work of a cell includes:

- active transport — moving ions and molecules across a membrane against a concentration gradient
- secretion — large molecules produced in some cells are exported by exocytosis
- endocytosis — bulk movement of large molecules and particles into cells
- biosynthesis — anabolic reactions producing large molecules (e.g. proteins from amino acids, cellulose from β-glucose)
- replication of DNA and synthesis of organelles — events during the cell cycle
- contraction of myofibrils — movement of actin filaments over myosin filaments
- activation of molecules — glucose is phosphorylated at the beginning of respiration

Exam tip

Energy cannot be created or destroyed. So *never* tell examiners that energy is produced. Respiration *releases* energy and *produces* ATP.

The energy for this work is made available from **adenosine triphosphate** (ATP). ATP is composed of adenine (a base) attached to a ribose (a pentose sugar) molecule, which is attached to a linear sequence of three phosphate groups (see Figure 1).

Exam tip

Both respiration and hydrolysis involve breakdown reactions, but you must distinguish between them. **Respiration** is the breakdown of organic molecules (e.g. glucose, fatty acids) so that energy is released to synthesise ATP; **hydrolysis** is the breakdown of large organic molecules (e.g. starch, lipids, proteins) into component molecules (e.g. glucose, fatty acids, amino acids) by the addition of water.

Knowledge check 1

Explain whether ATP is derived from a nucleotide of DNA or RNA.

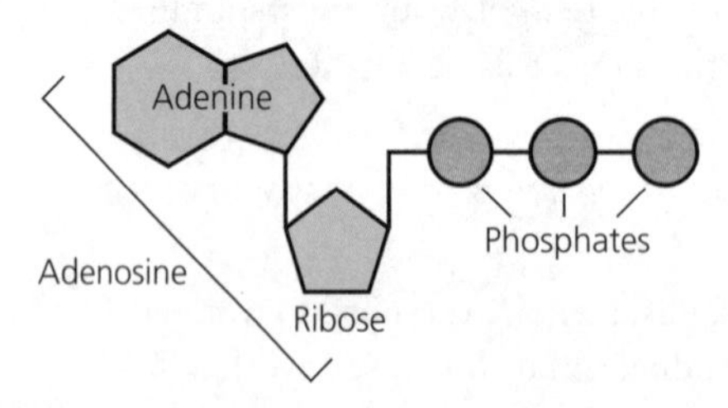

Figure 1 The structure of adenosine triphosphate (ATP)

Energy is released when ATP is hydrolysed to ADP and inorganic phosphate (P_i) by the enzyme ATPase. ATP is referred to as 'energy currency' because, like money, it can be used in different contexts and is constantly being recycled. At any one time there is only a small pool of ATP in the cell. When ATP is required it is recycled from ADP and P_i by the transfer of energy from the breakdown of glucose, fatty acids and, occasionally, amino acids. These are called **respiratory substrates**. Their breakdown, with the release of energy for ATP synthesis, is called **respiration**. The recycling of ATP is shown in Figure 2.

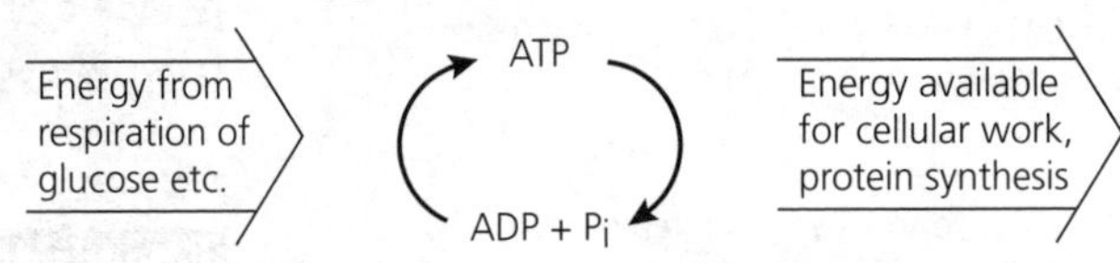

Figure 2 Recycling ATP: its hydrolysis with the release of energy and its synthesis as a result of respiration

Exam tip

Respiration occurs via a series of small steps. Some of these involve removal of hydrogen (**dehydrogenation**, an oxidation reaction), removal of carbon dioxide (**decarboxylation**) or addition of phosphate (**phosphorylation**).

During respiration, ATP is produced in two processes:

- **Substrate-level phosphorylation** — ATP is produced by the direct transfer of a phosphate group from a phosphorylated substance (a substance with a phosphate group attached) to ADP.
- **Oxidative phosphorylation** — ATP is produced from ADP and P_i as electrons are transferred along a series of carriers (the electron transport chain).

Two different forms of respiration are recognised:

- **Anaerobic respiration** does *not* require oxygen and can use glucose (carbohydrate) only. This is incompletely broken down and so only a little ATP is produced.
- **Aerobic respiration** requires oxygen and uses a variety of respiratory substrates that are completely broken down, producing a great deal of ATP.

Exam tip

Respiration includes oxidation and reduction. **Oxidation** involves addition of oxygen or removal of hydrogen (or electrons); **reduction** involves addition of hydrogen (or electrons). **Redox** reactions involve both oxidation and reduction: the molecule that accepts hydrogen is **red**uced; the molecule from which hydrogen is removed is **ox**idised.

Biochemistry of aerobic respiration

Aerobic respiration of glucose occurs in four stages in cells:

1. Glycolysis, which occurs in the cytoplasm.
2. Link reaction (pyruvate oxidation), which occurs in the mitochondrial matrix.
3. Krebs cycle, which occurs in the matrix of the mitochondrion.
4. Electron transport chain and oxidative phosphorylation, which occur across the inner membrane of the mitochondrion.

The **mitochondrion** is the organelle of aerobic respiration.

Glycolysis is the splitting of glucose in a metabolic pathway that has four major steps (see Figure 3). Glycolysis requires glucose, ATP, ADP, P_i and NAD (a coenzyme which picks up hydrogen). It produces reduced NAD (NADH) and ATP.

Exam tip

You should learn the steps of glycolysis so that you know where ATP is used, where NAD is reduced and where ATP is produced.

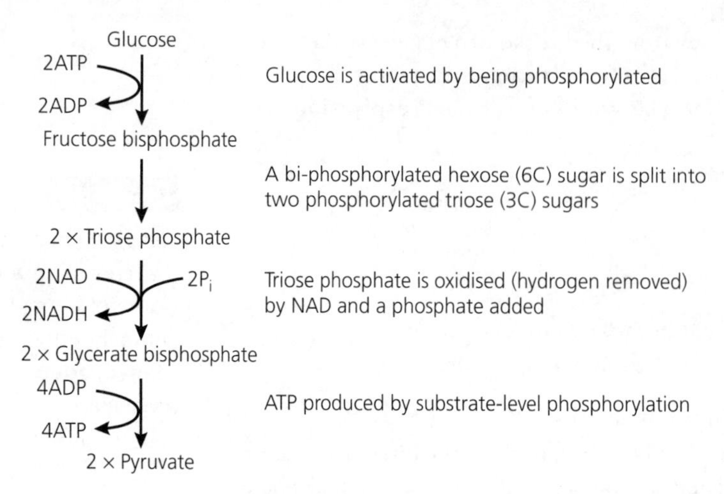

Figure 3 Glycolysis

For every molecule of glucose used in glycolysis, two molecules of ATP are used initially. However, four molecules of ATP are produced, so there is a net production of two ATP molecules per glucose molecule (this occurs in both aerobic and anaerobic respiration). During aerobic respiration, additional energy is available from reduced NAD (two molecules of reduced NAD are produced per glucose). For this additional energy to be made available, reduced NAD must pass into the mitochondrion to produce more ATP by oxidative phosphorylation. Pyruvate is transported into the mitochondrion where, if oxygen is available, it is further metabolised. Two molecules of pyruvate are produced per molecule of glucose. Therefore, when considering the breakdown of one molecule of glucose, the products of this further metabolism of pyruvate must be doubled.

In the mitochondrion, pyruvate is initially broken down in a **link reaction** (so called because it links with the Krebs cycle). The link reaction involves removal of hydrogen (dehydrogenation) and carbon dioxide (decarboxylation) from pyruvate, with the formation of an acetyl group (see Figure 4).

Carbon dioxide diffuses out of the mitochondrion and out of the cell. The hydrogen that is removed is picked up by NAD, producing NADH. The two-carbon acetyl group is carried by coenzyme A to produce acetyl coenzyme A.

Coenzymes work closely with the enzymes of respiration, carrying products of the breakdown to be used elsewhere:

- NAD carries hydrogen, as NADH, to be used in the electron transport chain (for oxidative phosphorylation).
- Coenzyme A carries the acetyl group, as acetyl coenzyme A, to be used in the Krebs cycle.

Exam tip

NAD is a coenzyme that accepts hydrogen atoms (and FAD, another coenzyme). Don't say that it accepts hydrogen *ions* or *molecules*.

Knowledge check 2

Use Figure 3 to explain the statement: 'The net yield of ATP is two molecules per molecule of glucose.'

Knowledge check 3

In the reaction from triose phosphate to glycerate bisphosphate (Figure 3), identify which molecule is oxidised and which is reduced.

Knowledge check 4

Which of the following are produced during glycolysis: NADH, ATP, NAD, lactate, CO_2, pyruvate?

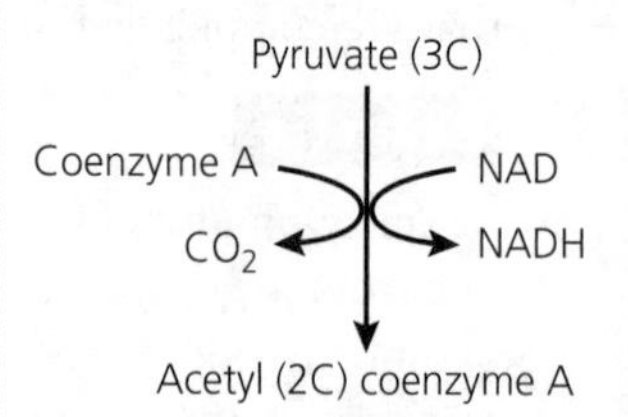

Figure 4 The link reaction (oxidation of pyruvate)

Knowledge check 5

State the role of the enzymes (a) pyruvate dehydrogenase, (b) pyruvate decarboxylase in the link reaction.

In the **Krebs cycle** the acetyl group, from acetyl coenzyme A, combines with a four-carbon acid to form a six-carbon acid. During each cycle:

- two steps involve removal of carbon dioxide molecules, which diffuse out of the mitochondrion and cell
- one step is a substrate-level phosphorylation, producing an ATP molecule
- four steps involve dehydrogenations, i.e. hydrogen is removed from the substrate

The Krebs cycle is shown in Figure 5.

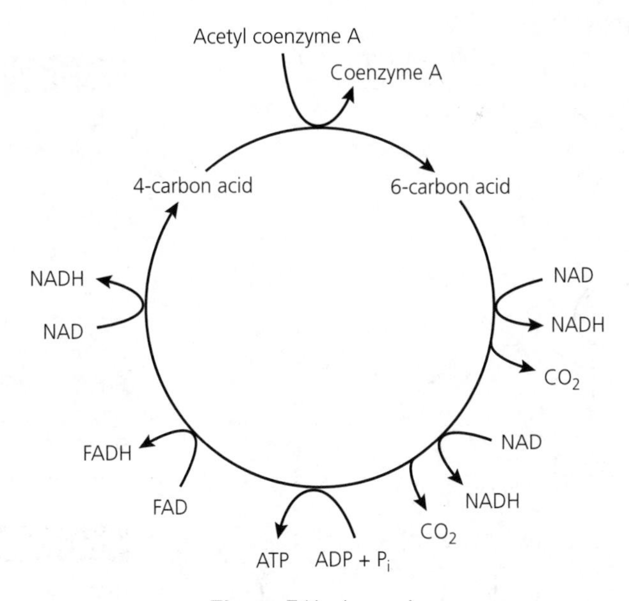

Figure 5 Krebs cycle

Removal of hydrogen is the most significant feature of the Krebs cycle. In three instances hydrogen is carried by NAD and in one case hydrogen is carried by FAD.

Hydrogens removed during glycolysis, the link reaction and the Krebs cycle are the source of electrons for the **electron transport chain**. In this chain, ATP is produced through **oxidative phosphorylation**. The process has the following features:

- Hydrogens are passed along initially and then subsequently only their electrons.
- Hydrogens, or their electrons, are passed through a series of carriers. NAD, flavoprotein and coenzyme Q act as hydrogen carriers; cytochromes and cytochrome oxidase are electron carriers.
- As a carrier receives hydrogens, or electrons, it is reduced. As it passes hydrogens or electrons along, it is oxidised. (This alternate reduction and oxidation of a carrier is called a redox reaction.)
- Oxygen acts as the final electron acceptor. It is reduced, forming water.
- The carriers lie at successively lower energy levels so that as electrons are transferred, energy becomes available.
- Sufficient energy becomes available at certain points in the chain to produce ATP through oxidative phosphorylation.
- From each NADH sufficient energy is released to produce three ATP molecules. Two ATP molecules are produced from each FADH (since hydrogen enters the

Knowledge check 6

How many molecules of (a) ATP and (b) NADH are produced in the Krebs cycle per molecule of glucose?

Exam tip

The Krebs cycle is the main source of reduced coenzymes, carrying hydrogen to the electron transport chain for the production of ATP.

Knowledge check 7

In a mammal, what happens to the CO_2 produced in the Krebs cycle?

chain from FADH at a point after the production of the first ATP). Therefore NADH has a greater yield of ATP.

- Electron transfer and oxidative phosphorylation are tightly coupled: one process takes place only if the other process can take place. For example, if ADP and P_i are not available for phosphorylation, then no electron transfer can take place and the coenzymes are all reduced, no matter how much oxygen is available.

The electron transport chain and oxidative phosphorylation are shown in Figure 6.

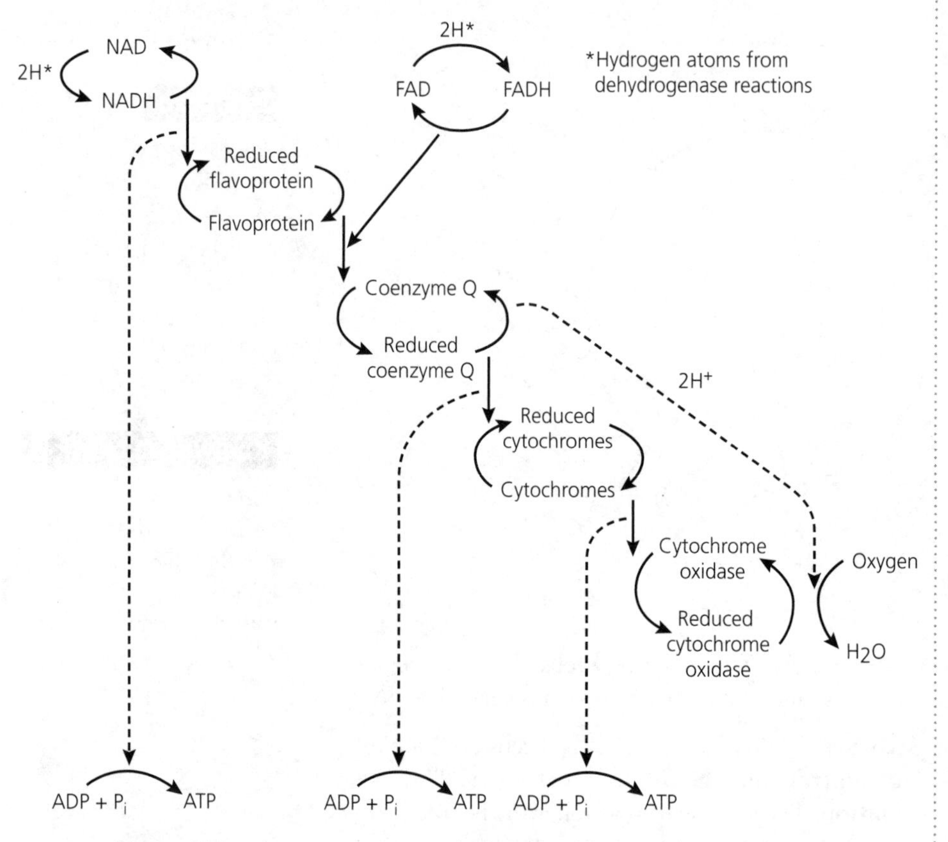

Figure 6 The electron transport chain and oxidative phosphorylation

Exam tip

Modern research supports the theory of chemiosmosis whereby electron transfer, through the carriers in the inner mitochondrial membrane, induces a proton pump which provides the energy for ATP synthesis. The number of ATPs subsequently produced differs from those traditionally given (as in Figure 6).

The total number of ATP molecules produced from aerobic breakdown of glucose is shown in Figure 7, which also serves as a useful overview of the process. Note the location of different stages: glycolysis in cytoplasm; link reaction and Krebs cycle in mitochondrial matrix; electron transport and oxidative phosphorylation in inner mitochondrial membrane.

Exam tip

The terms electron transport chain and respiratory chain are interchangeable.

Exam tip

A hydrogen atom consists of an electron and a proton. Only the hydrogen's electron passes along the cytochromes to finally reduce oxygen; the proton is released. The terms proton, hydrogen ion and H^+ all refer to the same particle.

Knowledge check 8

In precisely which part of the cell do (a) glycolysis, (b) the Krebs cycle, (c) the electron transport chain take place?

Knowledge check 9

Suggest how the availability of oxygen would limit the activity of (a) the electron transport chain and (b) the Krebs cycle.

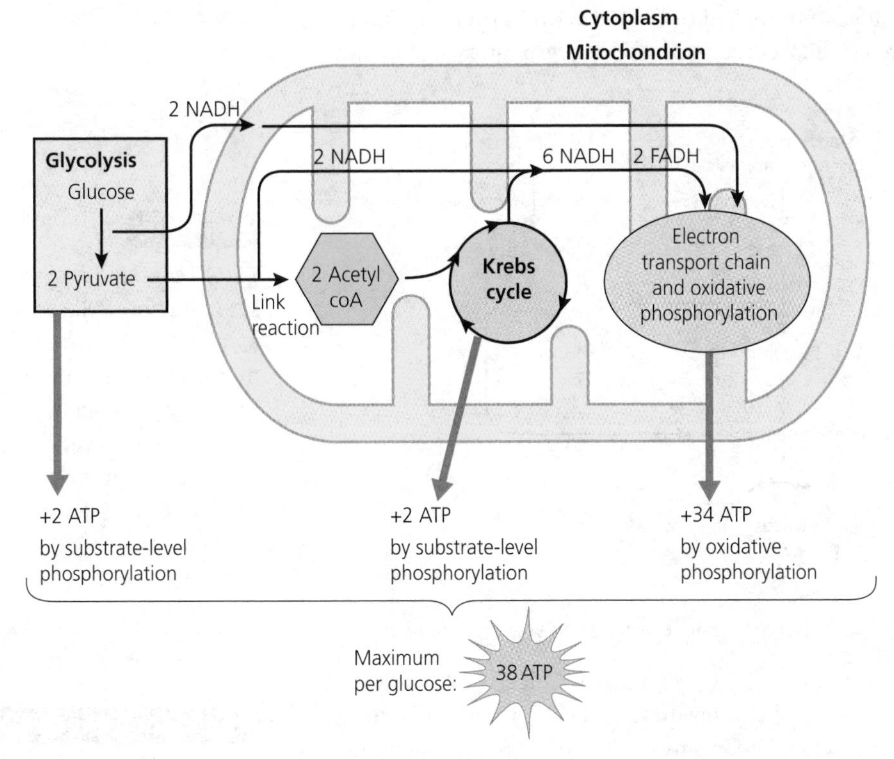

Figure 7 The total number of ATP molecules produced from the aerobic breakdown of one molecule of glucose

Figure 7 is based on the traditional view that three ATPs are produced from each reduced NAD and two ATPs are produced from each reduced FAD. With 10 NADH producing 30 ATPs and 2 FADH producing 4 ATPs, a total of 34 ATPs are synthesised from reduced coenzymes.

Knowledge check 10

The chemiosmotic theory has revealed that ten protons are pumped due to the electrons from NADH, six protons are pumped due to the electrons from FADH and four protons are needed to synthesise one ATP molecule. Calculate the number of ATP molecules produced due to each reduced coenzyme.

Knowledge check 11

What is the significance of the inner mitochondrial membrane being folded?

Other respiratory substrates

Respiratory substrates other than glucose can be respired aerobically.

Other **carbohydrates** are funnelled through the glycolytic pathway:

- Sugars such as fructose enter glycolysis.
- Polysaccharides such as starch (in plants) and glycogen (in animals) are hydrolysed into glucose or glucose phosphate and so enter glycolysis.

Triglycerides (lipids such as fats and oils) are hydrolysed into glycerol and fatty acids. Glycerol enters the glycolytic pathway. Fatty acids are energy rich, and ATP is generated in the mitochondrion either from initial splitting into two-carbon fragments (β-oxidation) or from the metabolism of the acetyl coenzyme A formed.

Proteins are hydrolysed to amino acids, which can be used to supply energy in dire situations such as starvation. The amino group is removed (deamination) and the remnant organic acid is converted into pyruvate or acetyl coenzyme A (mostly) or one of the acids in the Krebs cycle. (Excess dietary amino acids are converted to carbohydrates or triglycerides.)

Synoptic links

Biochemicals

In AS Unit 1 you learned about the structure of carbohydrates, lipids and proteins.

All substrates respired aerobically are funnelled through the Krebs cycle, often via acetyl coenzyme A, which is referred to as the focal point of respiratory metabolism. This is illustrated in Figure 8.

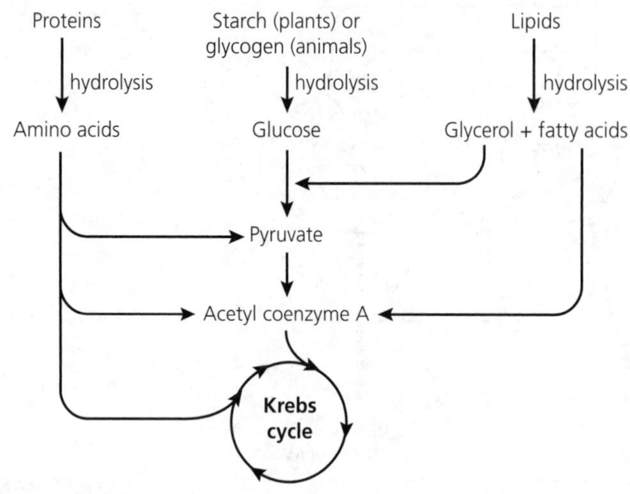

Figure 8 The respiratory metabolism of different respiratory substrates

Exam tip

Fatty acids and amino acids do not undergo glycolysis. They are metabolised and enter the aerobic respiration pathway mostly at acetyl coenzyme A. Unlike carbohydrates they can generate ATP *only* if oxygen is available.

Energy values of respiratory substrates are shown in Table 1. Since 1 g of lipid yields more than twice as much energy as 1 g of carbohydrate, only half the mass is needed to store an equivalent amount of energy. Organisms that have limited opportunities for obtaining food generally carry their 'energy store' as lipid. Examples include fats stored in camels' humps and in migrating birds such as ducks, and stores of oils in numerous types of plant seed.

Knowledge check 12

The fat in a camel's hump is a main energy reserve and weighs 30 kg. Using the data in Table 1, calculate the mass of carbohydrate that would provide the same amount of energy.

Table 1 Energy values of respiratory substrates

Respiratory substrate	Energy yield/kJ g^{-1}
Carbohydrate (polysaccharides and sugars)	16
Lipids (triglycerides)	39
Proteins (amino acids)	17

However, lipids can be respired aerobically only.

Biochemistry of anaerobic respiration

When oxygen is limited, the rate at which hydrogen carriers are oxidised (e.g. to NAD and FAD) is restricted and the cell may resort to anaerobic respiration (also referred to as fermentation). If no oxygen is available, there is no terminal electron acceptor so all the carriers in the hydrogen/electron transport chain remain reduced and the link reaction and the Krebs cycle cannot take place.

Anaerobic respiration is essentially glycolysis, with additional reactions to regenerate NAD. NAD is required for the oxidation of triose phosphate to glycerate bisphosphate (see Figure 3). In this reaction the NAD is converted into NADH, which cannot be converted back into NAD in the mitochondrion if oxygen is not available. Plants (and fungi) and animals have developed different pathways for regenerating NAD without the need for oxygen (see Figure 9). In both pathways, two molecules of ATP are produced from the metabolism of one molecule of glucose.

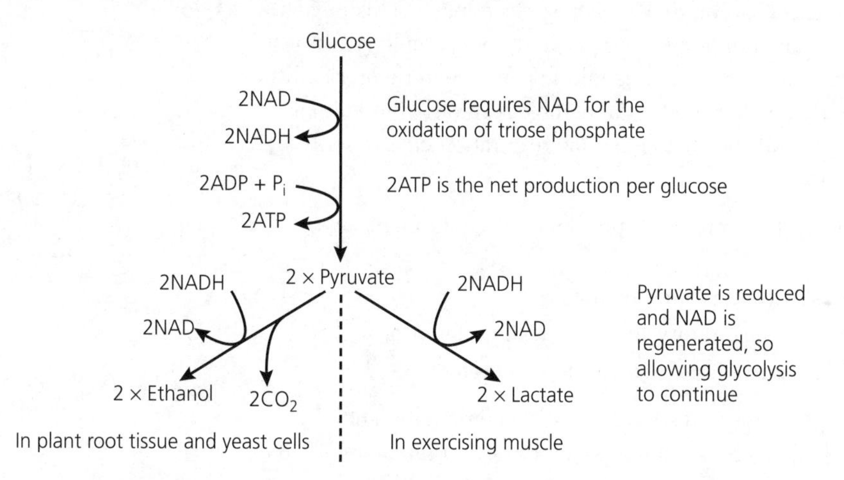

Figure 9 The anaerobic pathways of different organisms

The different pathways used by some plants (and fungi) and animals reflect different strategies:

- In plants and fungi, the anaerobic pathway allows the organism (e.g. yeast in suspension) or tissue (e.g. plant root in waterlogged conditions) to survive for a period of time without oxygen. The toxic effect of the waste products is reduced because they diffuse readily out of the organism — CO_2 is a one-carbon molecule and ethanol is a two-carbon molecule.
- In animals, the anaerobic pathway allows some tissues (e.g. mammalian muscle) to generate extra ATP over and above that generated aerobically. While only two molecules of ATP are produced per glucose in anaerobic respiration, this can occur extremely rapidly. It is this extra energy release that allows us to undertake strenuous exercise — for example, sprinting. The limiting factor is our tolerance to lactate, or rather lactic acid, which has a low pH and is toxic, so strenuous exercise can take place for only a brief period. After strenuous activity, the lactate in the muscle is carried in the blood to the liver; here, some of the lactate is used as a respiratory substrate to yield extra ATP for the conversion of the remaining lactate back to glucose (and subsequently to glycogen). The extra **recovery oxygen** required for the metabolism of lactate (and restoration of the oxymyoglobin levels) is supplied through continued rapid breathing and high heart rate — the '**oxygen debt**' is said to be paid back.

Respiratory quotients

The amount of carbon dioxide produced relative to the amount of oxygen consumed provides information about:

- the type of respiration taking place (i.e. aerobic or anaerobic)
- the nature of the respiratory substrate (e.g. carbohydrate or lipid)

Knowledge check 13

What is the difference between anaerobic respiration (fermentation) and glycolysis?

Exam tip

Remember that the purpose of the anaerobic pathways in mammals and yeast is to reoxidise NAD and allow glycolysis to continue, thereby generating ATP.

Knowledge check 14

Name the two products of anaerobic respiration in muscle tissue.

Knowledge check 15

Suggest how a build-up of lactic acid (lactate) leads to muscle fatigue in mammals.

Knowledge check 16

In animals, pyruvate can be regenerated from lactate when oxygen is available. Suggest why plant and yeast cells cannot regenerate pyruvate from ethanol.

The relative amount of carbon dioxide produced is called the **respiratory quotient** (RQ).

$$RQ = \frac{CO_2 \text{ produced}}{O_2 \text{ consumed}}$$

If the RQ is greater than 1, then extra carbon dioxide is being produced. This means that both aerobic and anaerobic respiration are taking place in, for example, yeast or plant tissue. (In yeast, if only anaerobic respiration is taking place, then theoretically the RQ value would be infinite. Remember too that anaerobic respiration in animal tissue does not produce carbon dioxide and that anaerobic respiration can use only carbohydrate as the respiratory substrate.)

If the RQ is less than or equal to 1, then aerobic respiration only is taking place, at least in yeast or plant tissue:

- If the RQ is 1, the respiratory substrate is carbohydrate.
- If the RQ is approximately 0.7, the respiratory substrate is lipid (fat or oil).
- If the RQ is approximately 0.9, the respiratory substrate is protein.

In general, however, a variety of respiratory substrates (mostly a combination of carbohydrate and lipid) is used. This is why, in humans, the RQ is in the region of 0.85.

Knowledge check 17

The following equation shows the aerobic respiration of stearic acid (a fatty acid):

$$C_{18}H_{36}O_2 + 26O_2 \rightarrow 18CO_2 + 18H_2O$$

Calculate the RQ for the respiration of stearic acid.

Practical work

Use a respirometer:

- Calculate oxygen uptake, carbon dioxide production and RQ values.

Use a redox indicator (e.g. methylene blue):

- Demonstrate the role of hydrogen acceptors.

Summary

- Respiration is the process by which organic molecules (mainly carbohydrates and lipids) are broken down, releasing energy to produce ATP.
- Aerobic respiration of glucose involves four stages:
  - In glycolysis, glucose is broken down in the cytoplasm to produce two molecules of pyruvate and two molecules of reduced NAD (NADH), with a net production of 2ATP by substrate-level phosphorylation; both pyruvate and NADH enter mitochondria.
  - The link reaction, in the mitochondrial matrix, converts pyruvate to acetyl coenzyme A, producing CO_2 and more NADH.
  - Acetyl coenzyme A is taken up by the Krebs cycle (also in the mitochondrial matrix), which generates CO_2 (twice), NADH (three times), FADH (once) and ATP (once, by substrate-level phosphorylation).
  - The reduced coenzymes (NADH and FADH) supply electrons to the electron transport chain (in the inner mitochondrial membrane) in which O_2 is the final electron acceptor. The transfer of electrons along carriers at lower energy levels is coupled to the synthesis of ATP, a process called oxidative phosphorylation. The coenzymes are reoxidised and so NAD (and FAD) are available to the dehydrogenase reactions of glycolysis, the link reaction and the Krebs cycle.

- Anaerobic respiration occurs without the use of oxygen. It involves glycolysis with additional reactions (producing lactate in animals and bacteria, and ethanol and CO_2 in plants and fungi) to regenerate NAD, so allowing continued production of ATP (even for a short period).
- Carbohydrate, lipid and protein can all be respired aerobically, with lipid yielding more energy per unit mass than carbohydrate. Only carbohydrate (glucose) can be respired anaerobically.
- The respiratory quotient or RQ (ratio of CO_2 produced to O_2 used) can be used to determine the type of respiration and respiratory substrate.

Photosynthesis

Photosynthesis is the process by which green plants, using energy from sunlight, produce a simple carbohydrate from the inorganic molecules carbon dioxide and water. Photosynthesis also releases oxygen, which is needed for aerobic respiration. The simple carbohydrate synthesised is used to produce other carbohydrates (such as starch, sucrose and cellulose), fatty acids and lipids, and amino acids and proteins. In doing this, plants make their own food — they are **autotrophic**.

The structure of a chloroplast is shown in Figure 10.

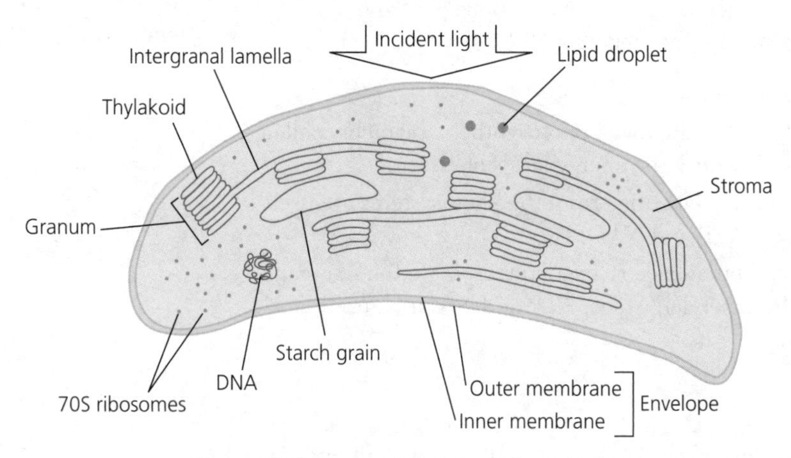

Figure 10 The structure of a chloroplast

Knowledge check 18

Name two other organelles that are surrounded by a double membrane (envelope).

Chloroplasts are orientated in the cell to ensure maximum exposure of the thylakoid surface to the direction of incident light.

Light is absorbed by a number of different pigments on the lamellae (e.g. thylakoids). There are two main classes of photosynthetic pigments:

- **chlorophylls** — chlorophyll *a* and chlorophyll *b*
- **carotenoids** — for example, β-carotene, the pigment in carrots

The **absorption spectrum** (Figure 11a) shows how much light a particular pigment absorbs at each wavelength. The chlorophylls absorb light in the blue–violet and red parts of the spectrum. The carotenoids absorb light from the blue–violet part of the spectrum. Different pigments absorb light of different wavelengths so the combined effect is

Absorption spectrum The amount of light absorbed by pigments at different wavelengths of light.

to increase the range over which light is effectively absorbed (see Figure 11a). The absorption spectrum can be compared with a graph of the rate of photosynthesis against wavelength, which is called an **action spectrum**, as shown in Figure 11(b). The shape of the action spectrum is similar to that of the absorption spectrum for the combined photosynthetic pigments — for example, the absorption spectrum shows that little green light is absorbed and therefore that green light is relatively ineffective in photosynthesis.

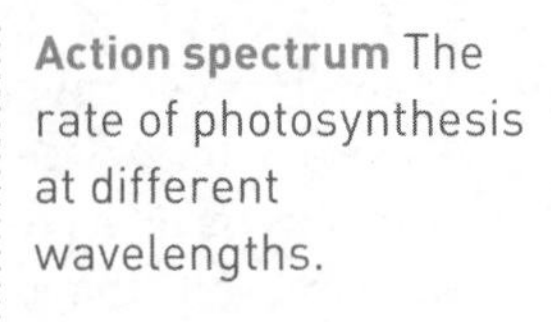

Action spectrum The rate of photosynthesis at different wavelengths.

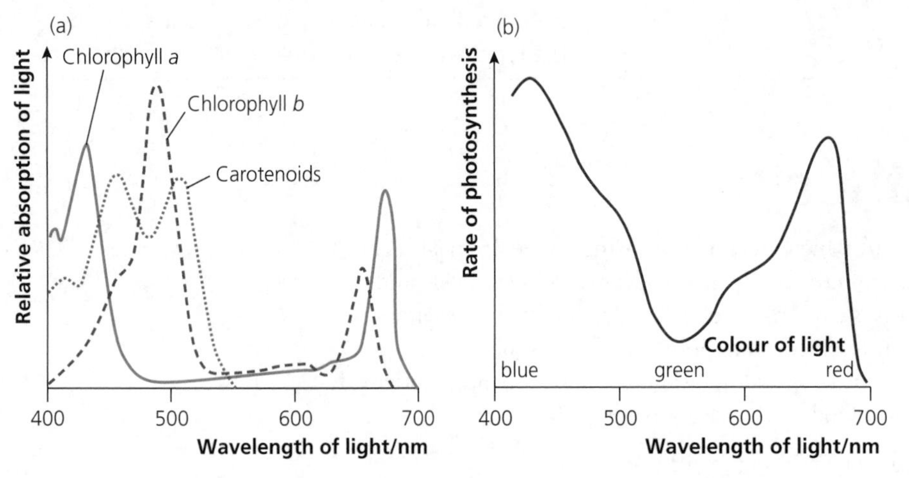

Figure 11 (a) The absorption spectrum for photosynthetic pigments. (b) The action spectrum for photosynthesis

Knowledge check 19

a Which area of the spectrum is not absorbed well by any of the photosynthetic pigments?

b What would be the 'best' colour for a photosynthetic pigment, allowing it to absorb light of all wavelengths?

There are three main stages to photosynthesis:

- Light harvesting — capturing of light by groups of photosynthetic pigments called photosystems. Light energy is used to raise the energy levels of electrons.
- Light-dependent reactions — energised electrons from the photosystems are used to produce energy-rich compounds, ATP and NADPH.
- Light-independent reactions — carbon dioxide is 'fixed' (incorporated) into organic form and the products of the light-dependent stage, ATP and NADPH, used to convert this to simple carbohydrate.

Light harvesting

Absorption of light is carried out by clusters of pigment molecules organised into photosynthetic units or **photosystems** located on the lamellae of chloroplasts. Each unit contains several hundred chlorophyll and carotenoid molecules. One particular chlorophyll *a* molecule, called a **primary pigment**, acts as a **reaction centre** for the photosystem. The remaining pigment molecules of the photosystem are called **accessory pigments** and represent an **antenna complex** absorbing light energy. Light energy absorbed by an accessory pigment creates an excitation energy that is passed along a chain of pigment molecules to the reaction centre. Resonance transfer of energy occurs, i.e. it is the energy that is transferred, not the electrons. Energy from many pigment molecules in the antenna complex is funnelled to the reaction centre. Energy reaching the reaction centre causes electrons in the primary chlorophyll *a* molecule to move up to a higher energy level. Such is the increased energy level of the electron that it is emitted by the chlorophyll *a* and taken up by an electron acceptor. Light harvesting is illustrated in Figure 12.

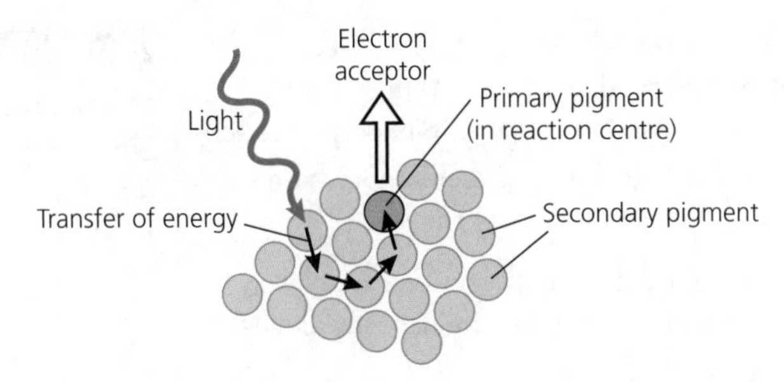

Figure 12 Light harvesting in a photosystem

There are two photosystems, photosystem I (PS I) and photosystem II (PS II):

- They differ in the proportions of different accessory pigments that make up the cluster.
- They differ in their primary pigments. PS I has a form of chlorophyll *a* that has a light absorption peak at 700 nm and is called P700. PS II has a form of chlorophyll *a* that has a light absorption peak at 680 nm and is called P680.

Light-dependent reactions

The excitation of the electrons that follows the harvesting of light by photosystems II and I results in an electron transfer pathway described as a **Z-scheme**. The sequence of events is shown in Figure 13.

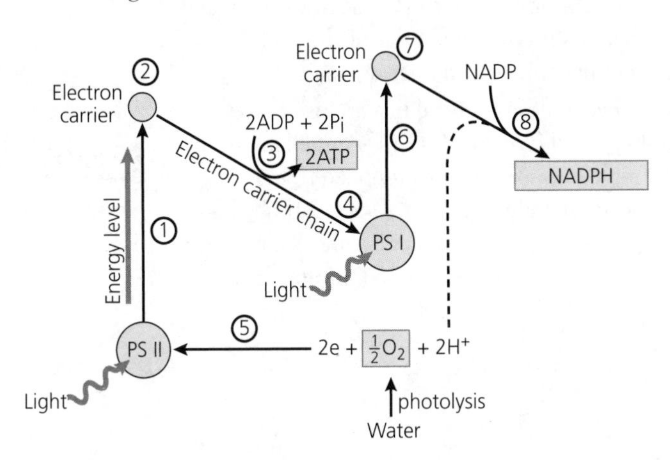

Figure 13 The light-dependent reactions of photosynthesis

(1) Light energy is trapped in PS II and boosts electrons to a higher energy level.

(2) Electrons are received by an electron acceptor.

(3) Electrons are passed from the electron acceptor along a series of electron carriers (cytochromes). Electron transfer is coupled to the synthesis of ATP (photophosphorylation).

(4) Electrons replace those lost in PS I.

(5) Electrons lost by PS II are removed from water molecules, the resultant photolysis of which produces oxygen and hydrogen ions.

(6) Light energy trapped by PS I boosts electrons to an even higher energy level.

(7) Electrons are removed by an electron acceptor.

(8) Hydrogen ions from the water combine with electrons from the second electron acceptor and these reduce NADP to NADPH.

Knowledge check 20

Explain why it is advantageous for photosystems to contain a complex of different pigments.

Exam tip

You must distinguish between 'dissociation of water' and 'photolysis of water': **dissociation of water** results in the formation of hydrogen ions (H^+) and hydroxyl ions (OH^-); **photolysis of water** results in electrons (that replace those 'lost' by chlorophyll *a* in PS II), hydrogen ions and oxygen (O_2).

Knowledge check 21

In the light-dependent reactions, what happens to electrons that come from (a) a chlorophyll molecule in photosystem I, (b) a chlorophyll molecule in photosystem II and (c) a water molecule?

Exam tip

While both NAD and NADP carry hydrogen, remember that NAD is used in respiration and NAD**P** in **p**hotosynthesis.

The most important consequences of this electron transfer are:

- synthesis of ATP via **photophosphorylation** and production of NADPH, both of which are used in the metabolic reactions whereby simple carbohydrate is synthesised
- release of oxygen as a waste product and its diffusion out of the chloroplast; any that is not used in respiration diffuses out of the cell and out of the leaf

Electron acceptors and carriers involved in the light-dependent reactions are closely associated with the photosystems and are located on the granal lamellae (thylakoids) of chloroplasts.

Knowledge check 22

Identify two similarities and two differences between photophosphorylation and oxidative phosphorylation.

Light-independent reactions: the Calvin cycle

The light-independent reactions (Calvin cycle) take place in the stroma of the chloroplasts. Four stages are recognised:

- *Fixation of carbon dioxide* — carbon dioxide enters the stroma and combines with the five-carbon compound **ribulose bisphosphate (RuBP)**. This reaction is catalysed by the enzyme **ribulose bisphosphate carboxylase** (commonly called **rubisco**). Two molecules of **glycerate phosphate (GP)** are formed.
- *Reduction of glycerate phosphate* — ATP and NADPH from the light-dependent reactions convert GP into **triose phosphate (TP)**. ATP provides the energy while NADPH provides the reducing power for this reaction. The ADP and NADP formed leave the stroma and enter the granal lamellae.
- *Regeneration of ribulose bisphosphate* — five out of every six TP molecules produced are used to regenerate ribulose bisphosphate using the remainder of the ATP from the light-dependent reactions as a source of phosphate and energy.
- *Product synthesis* — while for every six TP molecules produced five are used to regenerate RuBP, one TP represents 'profit'. The TP molecules so generated are used to produce a range of products — for example, hexose sugar is produced from two TP molecules. Other molecules are produced, including starch (storage carbohydrate from the hexose sugar), glycerol and fatty acids (and therefore lipids), and amino acids (and therefore proteins) with nitrogen being obtained from absorbed nitrate ions.

Knowledge check 23

Name three molecules involved in CO_2 fixation.

The light-independent reactions are shown in Figure 14.

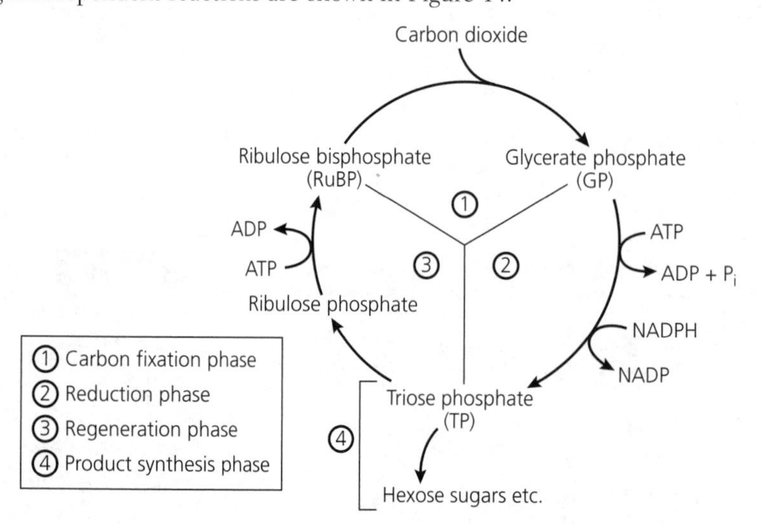

Figure 14 The light-independent reactions: the Calvin cycle

Knowledge check 24

Explain how reduced light intensity can affect the rate of the light-independent stage of photosynthesis.

Environmental factors that affect the rate of photosynthesis

The main factors that affect the rate of photosynthesis are light intensity, carbon dioxide concentration and temperature. These are called **limiting factors**. A limiting factor is that factor, of a number of possible factors, which is determining the rate at which the process is taking place. When a limiting factor is increased, the process takes place at a faster rate. Given that there is some carbon dioxide available and some warmth, light intensity is a limiting factor at low light intensities. This is illustrated in Figure 15.

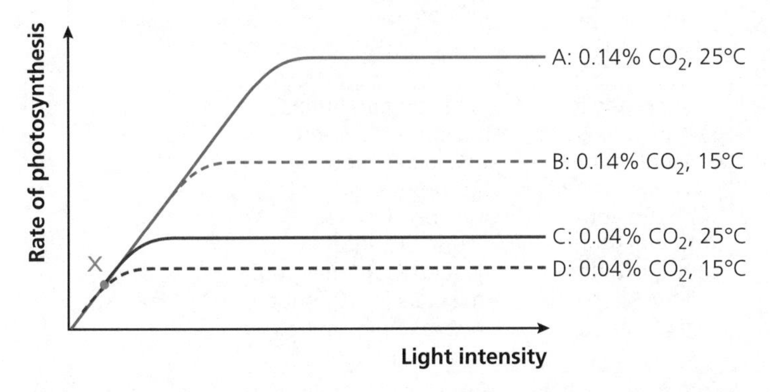

Figure 15 The effects of light intensity, carbon dioxide concentration and temperature on the rate of photosynthesis

Figure 15 is a complex graph showing the effects of light intensity, carbon dioxide concentration and temperature on the rate of photosynthesis.

- Light is not the limiting factor at higher light intensities (the plateau region of the graph). As light intensity is increased, the rate of photosynthesis is eventually limited by some other factor.
- Carbon dioxide is the limiting factor at high light intensity and over a range of temperatures:
  - At 15°C, the rate of photosynthesis increases significantly from D to B.
  - At 25°C, there is an even greater increase from C to A.
- Temperature appears to be an important limiting factor at high light intensity *and* at the higher carbon dioxide concentration. There is little increase in the rate of photosynthesis at the lower 0.04% carbon dioxide concentration (from D to C). At 0.14% carbon dioxide, the rate of photosynthesis increases significantly from B to A.

The above is a specific interpretation of Figure 15. A word of caution here — graphs must be interpreted according to the information available. For example, a graph comparing the rates of photosynthesis at 25°C and 5°C would probably show a significantly decreased rate at the lower temperature because the rate of enzyme action in the Calvin cycle would be greatly reduced.

Knowledge check 25

In Figure 15, which factor is limiting the rate of photosynthesis at point X?

Knowledge check 26

Explain why, in Figure 15, the rate of photosynthesis is higher in curve A than in curve C.

The effects of the three factors are further discussed in Table 2.

Table 2 The effects of environmental factors on the rate of photosynthesis

Environmental factor	Reason for influence	Effect of factor	Effect in the environment
Light intensity	Greater light intensity increases energy available for the light-dependent reactions and, therefore, production of ATP and NADPH for the light-independent reactions	As light intensity increases, the rate of photosynthesis increases proportionally, until it is limited by some other factor	On a winter's day the light intensity may be limiting (if it is not too cold); short day length, during which photosynthesis is possible, is a major influence
Carbon dioxide concentration	An increase in CO_2 concentration increases the carboxylation of RuBP and, therefore, production of GP	An increase in carbon dioxide concentration increases the rate of photosynthesis, until it is limited by another factor	In bright conditions during the summer, CO_2 is probably limiting; levels of CO_2 may be increased artificially in a greenhouse; the global increase in CO_2 should increase the rate of photosynthesis
Temperature	Enzymes catalyse the light-independent reactions; as the enzymes approach their optimum temperature, activity increases; above optimum temperature enzymes are denatured	Temperature is a limiting factor only at very low or very high temperatures or when there are relatively high levels of light and CO_2	On a very cold winter's day, temperature may be limiting; in a greenhouse, additional heating may be provided during the winter; a temperature of around 25°C is optimal; plants survive high summer temperatures because transpiration has a cooling effect

Practical work

Paper chromatography of plant pigments:

- preparation and running of the chromatogram
- calculation of R_f values

Use a redox indicator (e.g. DCPIP) to demonstrate the role of hydrogen acceptors.

Knowledge check 27

Explain why burning an oil-fired stove increases the growth of plants in a greenhouse.

Summary

- During photosynthesis, light energy is trapped and used to synthesise organic compounds from CO_2 and water; the chloroplast is the organelle involved in photosynthesis.
- During light harvesting, different chlorophyll and carotenoid molecules (clustered as photosystems on the lamellae) absorb light maximally at different wavelengths (as shown by the absorption spectrum). This allows photosynthesis to take place over a range of wavelengths (the action spectrum). As pigment molecules are energised, energy passes by resonance to a primary pigment that emits high-energy electrons.
- In the light-dependent stage, electrons pass from PS II to PS I through an electron transport system in the lamellae that synthesises ATP (photophosphorylation). PS II regains electrons from the photolysis of water; oxygen is released. Energised electrons from PS I reduce NADP to NADPH.
- In the light-independent stage (Calvin cycle), which occurs in the stroma, CO_2 is fixed by RuBP, forming glycerate phosphate. Using energy from ATP, this is reduced by NADPH to triose phosphate. Some of the triose phosphate, with additional ATP, is used to regenerate RuBP; some triose phosphate is converted into glucose, disaccharides, polysaccharides, lipids and proteins (during product synthesis).
- The rate of photosynthesis is affected by a number of factors, including light intensity, carbon dioxide concentration and temperature; the rate is determined by the limiting factor.

DNA as the genetic code

A **gene** is a length of DNA that acts as a coding device for the synthesis of a particular polypeptide. More precisely, it is a sequence of nucleotide bases that determines the order of the amino acids in the polypeptide produced.

Synoptic links

Nucleic acids

In AS Unit 1 you learned about the structure of nucleotides and nucleic acids, about polypeptides and proteins, and about ribosomes as the site of polypeptide synthesis.

It is the sequence of nucleotide bases that represents the **genetic code**. The main features of the genetic code are as follows:

- It is a **three-base (triplet) code** — a sequence of three nucleotide bases codes for an amino acid. There are four bases arranged in groups of three, so the number of different sequences is 4^3 or 64. As there are only 20 amino acids used for protein synthesis, this is more than enough sequences. (Note that a one-base code would only code for four amino acids and a two-base code would only code for 4^2 or 16 amino acids.)
- It is a **degenerate code**. Most amino acids have more than one triplet code, since there are more triplet codes (64) than amino acids (20). For example, the amino acid leucine has six codes and many have four (see Table 3, p. 24).
- It is a **non-overlapping code**. The triplet codes are read separately, i.e. each base is read only once. For example, AGAGCG is read as AGA for one amino acid and GCG for the next, rather than AGA, GAG, AGC, GCG.

Knowledge check 28

Explain why the genetic code must be triplet and not doublet.

Knowledge check 29

The gene for one of the polypeptides in haemoglobin consists of 438 bases. How many amino acids are in the polypeptide?

DNA does not act directly in protein synthesis. It is transcribed into a single-stranded RNA molecule that acts as a messenger. There are a number of reasons for this:

- DNA is a very long molecule (so long that in eukaryotes, each is supercoiled into a chromosome during nuclear division). It is too long to move out of the nucleus into the cytoplasm where protein synthesis takes place.
- DNA is retained in the nucleus where, as the molecule of inheritance, it is better protected from the possibility of damaging changes.
- DNA is used to produce multiple copies of messenger RNA and so many polypeptides can be synthesised simultaneously.

The result is that there are two stages to protein synthesis:

- The genetic code in the DNA molecule in the nucleus is first copied into a molecule of messenger RNA (mRNA) in the process of **transcription**.
- The mRNA molecule then moves out of the nucleus onto a ribosome where its code is used to direct the synthesis of a polypeptide in the process of **translation**.

Exam tip

Many students confuse transcription and translation. It may help to remember that to 'scribe' is to *copy* (the genetic code) and that to 'translate' is to *put into a different language* (nucleotide sequence to amino acid sequence).

Protein synthesis: transcription

An overview of protein synthesis is given in Figure 16.

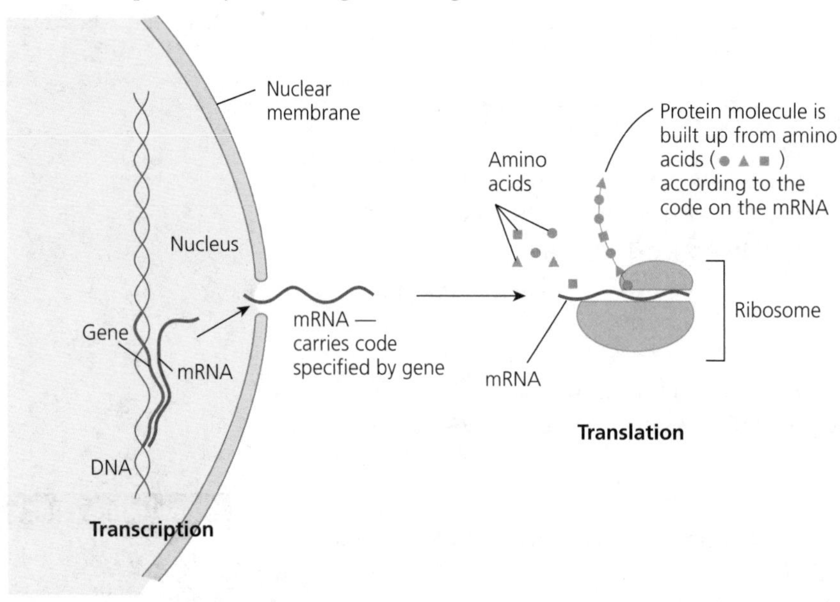

Figure 16 An overview of protein synthesis

The process of transcription (Figure 17) goes through the following sequence:

- The two strands of DNA are separated by the enzyme **helicase** using energy from ATP.
- The enzyme **RNA polymerase** binds to a region of the DNA near the beginning of the gene that is going to be transcribed.
- One of the exposed strands — the **template strand** — is used as a template for mRNA production. As RNA polymerase moves along the template strand, ribonucleotides are taken from the nucleoplasm and matched up by a process of complementary base pairing.
- **Complementary base pairing** involves the bases in the ribonucleotides forming hydrogen bonds with the exposed bases on the DNA template strand, according

Exam tip

The DNA (and mRNA) is read from the 3′ end to its 5′ end. Remember that nucleic acids have different ends.

to base pairing rules: a ribonucleotide with U binds with A in the DNA strand, G with C, A with T and C with G. The RNA polymerase binds the newly arrived ribonucleotides along their sugar–phosphate backbone.

- As the RNA polymerase moves along, the DNA double helix reforms. When the enzyme reaches the end of the gene it releases the fully formed **mRNA**. The mRNA produced is complementary to the nucleotide base sequence on the template strand of the DNA.

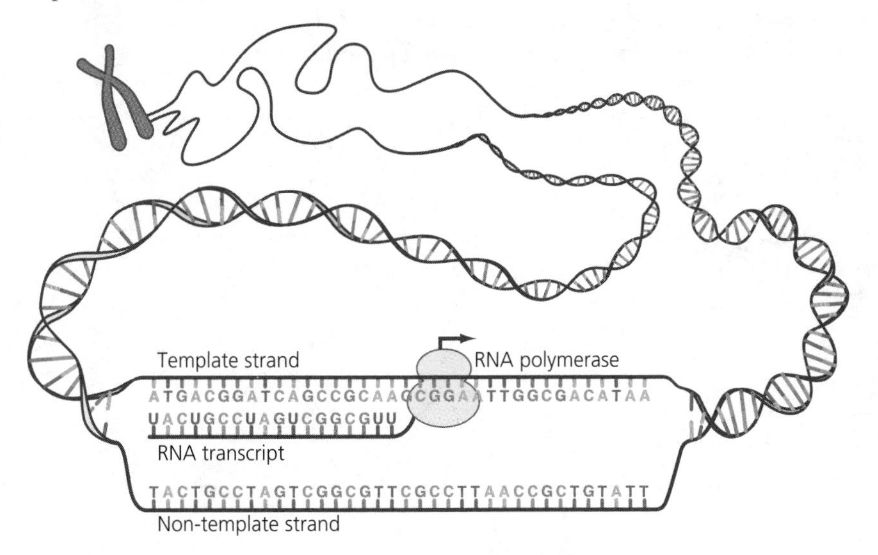

Figure 17 Transcription

In eukaryotic cells, the first mRNA formed (pre-mRNA) contains coding regions (**exons**) interspersed with non-coding regions (**introns**). To convert pre-mRNA into mRNA, the introns are cut out and the remaining exons joined or **spliced** together. The final version of mRNA, which now contains only the coding from the exons, passes out of the nucleus through a pore in the nuclear envelope. It moves to a ribosome, where the synthesis of the protein takes place. The production of functional mRNA is shown in Figure 18.

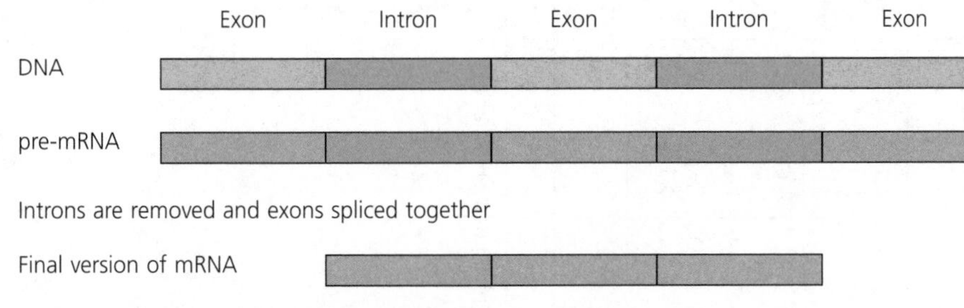

Figure 18 Production of functional mRNA — cutting out introns and splicing together the remaining exons

Exam tip

Following the removal of introns, the exons can be spliced in a range of combinations. This alternative splicing allows a gene to code for a number of different polypeptides. This explains how the 21 000 genes in humans can code for the 100 000 or more different proteins estimated to occur in the body.

Exam tip

Remember the pairing is A–T and G–C between DNA strands (notice A and T are angular letters, while G and C are curved letters). In RNA, remember to replace T with U.

Knowledge check 30

Give two similarities and two differences between transcription and DNA replication.

Knowledge check 31

If three exons (E1, E2 and E3) are spliced back together in alternative ways, what is the maximum number of polypeptides that can be produced? Explain your answer.

Exam tip

While eukaryotic genes contain introns, bacteria (prokaryotes) do not have introns in their genes and lack the enzymes to remove them. This means that if eukaryotic genes are to be functional in bacteria they must be inserted without introns — an important consideration in genetically engineering bacteria.

Protein synthesis: translation

The components of translation are:

- mRNA, containing a sequence of bases, triplets of which are called **codons**
- **ribosomes**, each of which consists of two subunits, one small and one large; a ribosome has two sites for mRNA attachment, a peptidyl (P) site and an aminoacyl (A) site
- **transfer RNA** (tRNA) molecules, each of which carries a specific amino acid at one end and a triplet base code, called an **anticodon**, at the other (see Figure 19)

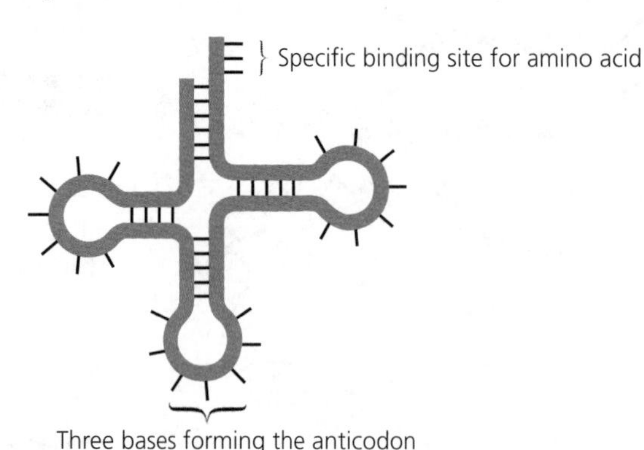

Figure 19 The structure of transfer RNA

The mRNA genetic codes (codons) for amino acids are shown in Table 3. The different amino acids are represented by three-letter abbreviations.

Table 3 The mRNA genetic codes (codons) for amino acids

First position (5′ end)	Second position: U		Second position: C		Second position: A		Second position: G		Third position (3′ end)
U	UUU	Phe	UCU	Ser	UAU	Tyr	UGU	Cys	U
U	UUC	Phe	UCC	Ser	UAC	Tyr	UGC	Cys	C
U	UUA	Leu	UCA	Ser	UAA	stop	UGA	stop	A
U	UUG	Leu	UCG	Ser	UAG	stop	UGG	Trp	G
C	CUU	Leu	CCU	Pro	CAU	His	CGU	Arg	U
C	CUC	Leu	CCC	Pro	CAC	His	CGC	Arg	C
C	CUA	Leu	CCA	Pro	CAA	Gln	CGA	Arg	A
C	CUG	Leu	CCG	Pro	CAG	Gln	CGG	Arg	G
A	AUU	Ile	ACU	Thr	AAU	Asn	AGU	Ser	U
A	AUC	Ile	ACC	Thr	AAC	Asn	AGC	Ser	C
A	AUA	Ile	ACA	Thr	AAA	Lys	AGA	Arg	A
A	AUG	Met	ACG	Thr	AAG	Lys	AGG	Arg	G
G	GUU	Val	GCU	Ala	GAU	Asp	GGU	Gly	U
G	GUC	Val	GCC	Ala	GAC	Asp	GGC	Gly	C
G	GUA	Val	GCA	Ala	GAA	Glu	GGA	Gly	A
G	GUG	Val	GCG	Ala	GAG	Glu	GGG	Gly	G

Knowledge check 32

Using the information in Table 3, determine which amino acid has the codon ACG.

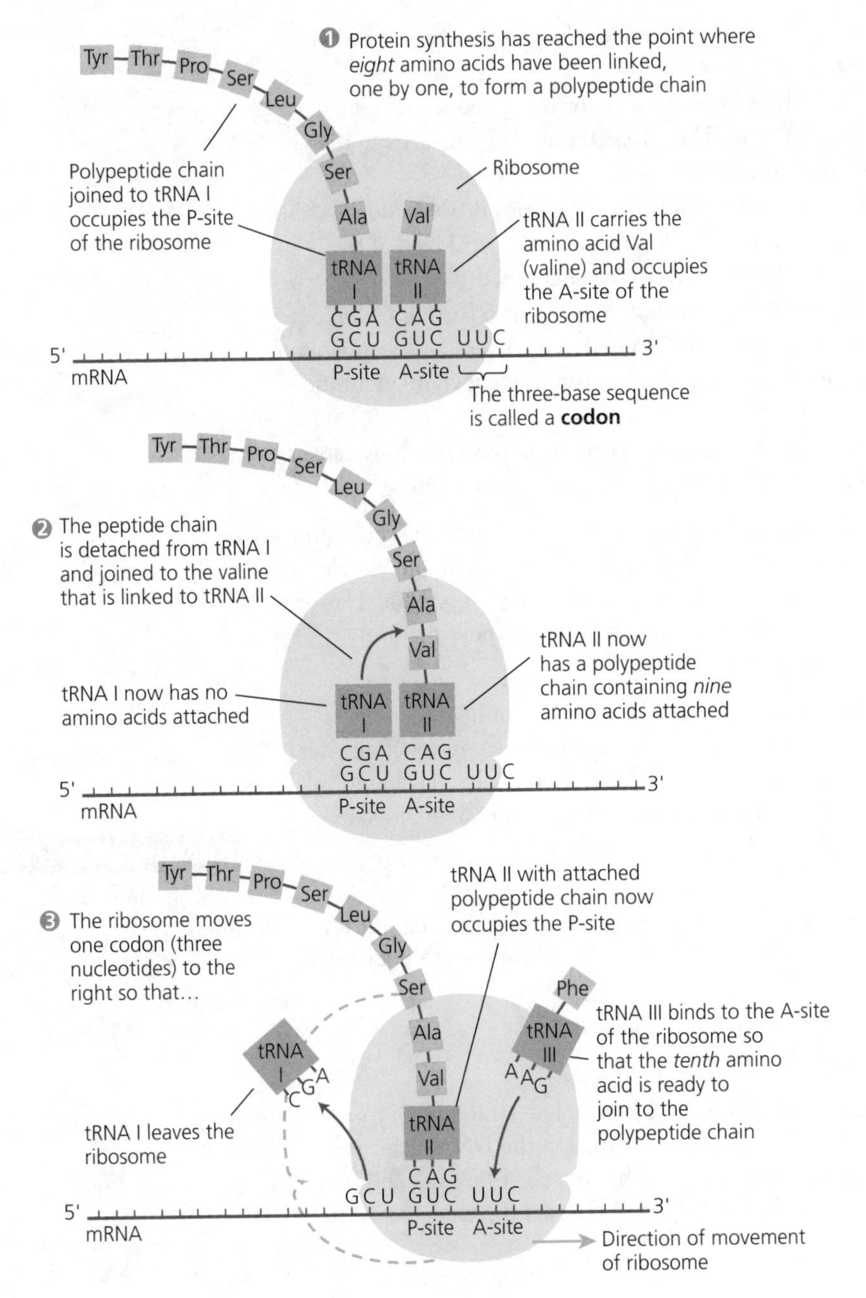

Figure 20 Translation

The process of translation (Figure 20) follows this sequence:

- The first two codons of the mRNA enter a ribosome, the first at the peptidyl (P) site and the second at the aminoacyl (A) site. The first codon is a start (initiation) codon (and is always AUG).
- Transfer RNA molecules (with specific amino acids attached) that have complementary anticodons to the first two codons of the mRNA bind to those codons.

Knowledge check 33

The sequence of triplet base codes on a template strand of DNA is CAGACATTC. Using the information in Table 3, determine: (a) the sequence of bases on the mRNA, (b) the tRNA anticodon for each mRNA codon, and (c) the sequence of amino acids (abbreviated names) in the polypeptide.

Knowledge check 34

Name: (a) the molecule that carries an amino acid, (b) the organelle on which translation occurs, and (c) the bond that forms between amino acids.

- A peptide bond, requiring energy from an ATP molecule, is formed between the amino acids carried by these two tRNA molecules.
- The ribosome moves along the mRNA by one codon, bringing the third codon onto the aminoacyl site on the ribosome. The tRNA that is freed returns to the cytoplasm where it will pick up another amino acid.
- A tRNA with a complementary anticodon binds with the third mRNA codon, bringing its amino acid into position next to the second amino acid held at the peptidyl site. The amino acids bond and the chain is lengthened.
- The ribosome moves along the mRNA by one codon, bringing a further codon onto the aminoacyl site. An amino acid carrying tRNA with a complementary anticodon binds with the codon and the amino acid bonds with the existing chain of amino acids held at the peptidyl site.
- The process is repeated until a stop or termination code is in position. This causes the polypeptide to be released. Translation of a polypeptide is complete.

In eukaryotic cells, the ribosomes are usually associated with endoplasmic reticulum (ER), forming rough ER. Newly synthesised polypeptides enter the ER and vesicles containing the polypeptides are 'nipped' off and move to the Golgi apparatus. Here they fuse with the formative face. In the Golgi apparatus the polypeptides are processed to produce final functional proteins.

These proteins include cell surface proteins, channel proteins, antibodies, some hormones (e.g. insulin) and enzymes. Enzymes are involved in the control of all metabolic pathways and thus in the synthesis of all non-protein molecules found in cells. Whatever their role, proteins determine the characteristics of an organism.

Gene mutation

A **gene mutation** is a spontaneous change in the DNA molecule. It occurs most often as a result of a mistake when DNA is replicating. Two examples are nucleotide base substitutions and deletions:

- In a substitution, one base is replaced by another base.
- In a deletion, one base is omitted altogether.

Knowledge check 35

Explain how the degenerate nature of the genetic code reduces the effects of base substitutions.

In the following **base substitution**, a thymine on the template strand of DNA is replaced (mistakenly) by an adenine. The result is a change in the DNA code, a change in the mRNA codon for the amino acid brought onto the ribosome and so a change in the amino acid sequence of the polypeptide being synthesised.

Original template strand:

C–T–C—C–T–T—T–T–T… → glutamic acid — glutamic acid — lysine…

Altered template strand:

C–A–C—C–T–T—T–T–T… → valine — glutamic acid — lysine…

The amino acid valine replaces the amino acid glutamic acid in the polypeptide. In fact, this is the mutation that results in the change in the β-chain of haemoglobin which causes sickle-cell anaemia. The presence of a different amino acid in the polypeptide changes the structure of the haemoglobin — it is defective and causes changes in the shape of red blood cells which, consequently, may clog blood vessels, thus depriving vital organs of their full supply of oxygen.

Note that substitutions of the third base in the triplet codes may not affect the amino acid coded for because of the degenerate nature of the code (see Table 3).

In the following **base deletion**, guanine is lost from the template strand so that all the other bases move up by one. This is called a **frameshift**. The change affects not just one triplet code but all those thereafter. A multiple change in the amino acid sequence occurs and the final amino acid is missing.

Original template strand:

T–A–**G**—T–G–A—A–A–C—G–G–C... → isoleucine — threonine — leucine — proline

Altered template strand:

T–A–T—G–A–A—A–C–G—G–C... → isoleucine — leucine — cysteine

In the above example, the first triplet code is changed at the third base without affecting the amino acid it codes for. However, with a different amino acid sequence thereafter, the polypeptide produced would fold to form a different shape. If an enzyme was being produced it could lack the precise shape of the active site necessary for it to function. There are many examples of deletion mutations. For example, cystic fibrosis is caused by a deletion mutation, though in this case three bases are deleted so that one amino acid is missing. The effect is a non-functional membrane protein.

Knowledge check 36

Explain why a base substitution does not cause a 'frameshift'.

Exam tip

If asked about how a mutation may lead to a change in the encoded protein you must be precise. The change in base sequence in the DNA leads to a change in mRNA which leads to a change in the amino acid sequence of the polypeptide which leads to different globular conformation of the protein.

Gene mutations, occurring spontaneously and randomly, can be inherited, usually from carrier parents. A gene mutation in an organ that gives rise to gametes (e.g. the ovary) will be passed on to offspring without any family history of the genetic disorder.

Mutation in a body cell (called a somatic mutation) may alter the way the cell functions and, indeed, may lead to its death. Alterations of certain genes may make the cell divide uncontrollably, producing a tumour that may become malignant or cancerous.

Epigenetics

Epigenetics is the study of changes in gene expression that do not involve changes in the base sequence of the DNA (the genetic code). They are caused by changes in the cell environment so that some genes are 'switched on' while others are 'switched off'.

DNA methylation

Methyl groups can attach to particular bases of the DNA, usually cytosine (Figure 21). These chemical 'tags' are altered throughout life in response to biochemical signals in the cell (which may in turn be influenced by external factors such as diet). The methylation of DNA prevents the attachment of RNA polymerase and so inhibits transcription. Methylation therefore prevents the gene being expressed — the gene is said to be 'silenced'.

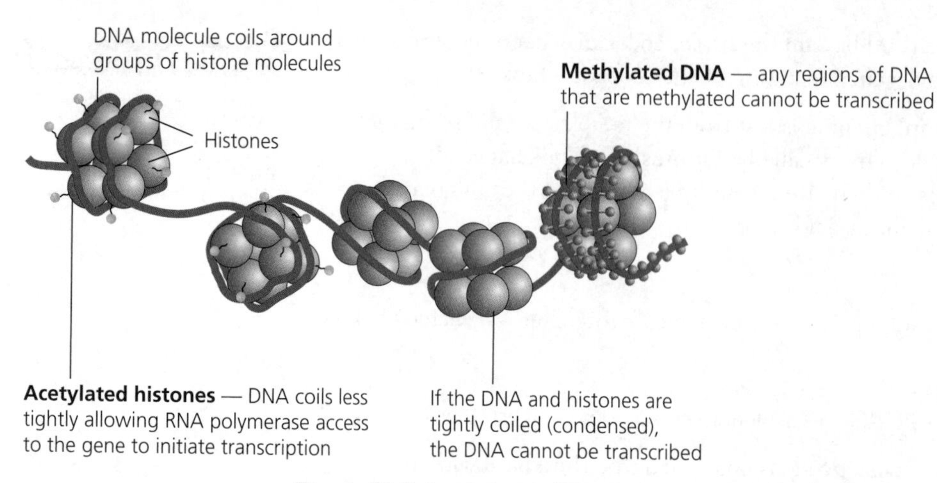

Figure 21 Epigenetic modifications

Histone modification

In a eukaryotic chromosome, DNA is closely associated with proteins called histones. DNA is wound round the histones, producing the chromatin that is visible in the nucleus of a non-dividing cell. When the DNA–histone complex is tightly condensed (heterochromatin), the DNA cannot be transcribed. The histones can be altered by adding acetyl, methyl or phosphate groups, which affects whether the DNA is condensed or not. For example, when acetyl groups are added to histones, the DNA is wound less tightly round the proteins, making it easier for RNA polymerase to bind to the DNA; removing acetyl groups from the histones makes the DNA wind more tightly round the histones, so transcription is suppressed (Figure 21).

Significance of epigenetic modification

Epigenetic modifications occur during embryonic development, when cells are differentiating. This ensures that some genes are switched off while others are switched on, so that only particular sets of genes are expressed in specific types of specialised cell.

Ageing and age-related diseases are associated with epigenetic changes. During the ageing process, some genes that are active in youth become inactive and other genes that were suppressed during early development become activated. The potential reversibility of these epigenetic changes may allow the ageing process and the progress of age-related diseases to be altered.

Epigenetic changes may be involved in cancer development. They can cause **tumour suppressor genes** (e.g. gene *p*53) to be silenced, or **oncogenes** to be activated. Drugs are being developed to treat such cancers by reversing epigenetic changes, e.g. by removing acetyl 'tags' on histones (silencing genes) or removing methyl groups on DNA (activating genes).

Epigenetic modifications are long-lasting and are inheritable. They can be passed on during mitotic division and from parent to offspring.

Knowledge check 37

Explain why monozygotic twins, whose DNA is identical at fertilisation, become increasingly different from each other as they get older.

Tumour suppressor gene Gene that protects a cell from uncontrolled division.

Oncogene Gene that can lead to uncontrolled cell division.

Knowledge check 38

DNA methylation has been shown to decrease with age. This decrease in methylation may cause a cell to become cancerous through its effect on which type of gene, tumour suppressor gene or oncogene? Explain your answer.

Summary

- A gene is a length of DNA with a base sequence that codes for the synthesis of a particular polypeptide. The genetic code is a triplet, degenerate, non-overlapping code.
- In the nucleus, the DNA code is transcribed onto many mRNA molecules that move through pores in the nuclear envelope into the cytoplasm to be translated into amino acid sequences.
- During transcription, the strands of DNA are separated along the length of the gene; one strand acts as the template for the assembly of complementary mRNA, catalysed by RNA polymerase; introns are removed and exons spliced together to form functional mRNA.
- During translation, the first two codons of the mRNA enter a ribosome; tRNA molecules with complementary anticodons bind with the RNA codons; the specific amino acids carried onto the ribosomes are joined by peptide bonding; the ribosome moves along the mRNA one codon length, the codon is translated and the process is repeated until the polypeptide is complete.
- Gene (point) mutations are spontaneous changes in a single base on the DNA molecule: substitution replaces one base with another and causes an amino acid substitution in the polypeptide formed (or not since the code is degenerate); deletion removes a base, causing a frameshift (altering all triplet codes after the point of mutation).
- Changes in gene expression also occur through epigenetic modifications (which do not alter the genetic code): DNA methylation inhibits gene expression ('silences' genes); histone acetylation activates gene expression.

Gene technology

Gene technology involves the manipulation of DNA and includes DNA amplification, genetic fingerprinting, gene sequencing, gene testing, pharmacogenetics, genetic modification of organisms, gene therapy and knockout gene technology.

The polymerase chain reaction (PCR)

The **polymerase chain reaction** (**PCR**) is used to *amplify target DNA sequences* that are present in a DNA source. While it mimics the natural process of DNA replication, it can generate billions of copies of a DNA sample within a few hours and so has become fundamental to many aspects of gene technology.

Synoptic links

DNA replication

In AS Unit 1 you learned about the natural process of DNA replication and also about heat-stable enzymes.

The PCR requires:

- a DNA sample that includes the selected region for replication
- the synthesis of **primers** — short strands of DNA (of about 20 nucleotides) that are complementary to the sequence at the start of each strand of the region to be amplified

- the enzyme DNA polymerase, extracted from thermophilic bacteria and which is therefore thermostable (heat-stable)
- free deoxyribonucleotides

The mixture is placed in a thermal cycler, the automated machine that changes the temperature at which the reagents are incubated in a pre-programmed sequence (Figure 22):

- The DNA to be amplified is heated to 95°C for 1 minute. This breaks hydrogen bonds between the paired bases and separates the two strands.
- The mixture is cooled to 50–60°C, which allows primers to **anneal** to the start of each strand of the complementary DNA region. The primers prevent the DNA strands rejoining and act as signals to the polymerase enzymes to start adding nucleotides.
- The mixture is heated to 72°C for 1–2 minutes and the thermostable polymerase enzyme (at its optimum temperature) copies each strand, starting at the primers.
- The process is repeated and with each cycle the number of DNA molecules is doubled — a chain of 20 cycles would produce millions of copies of the original DNA because of this exponential amplification.

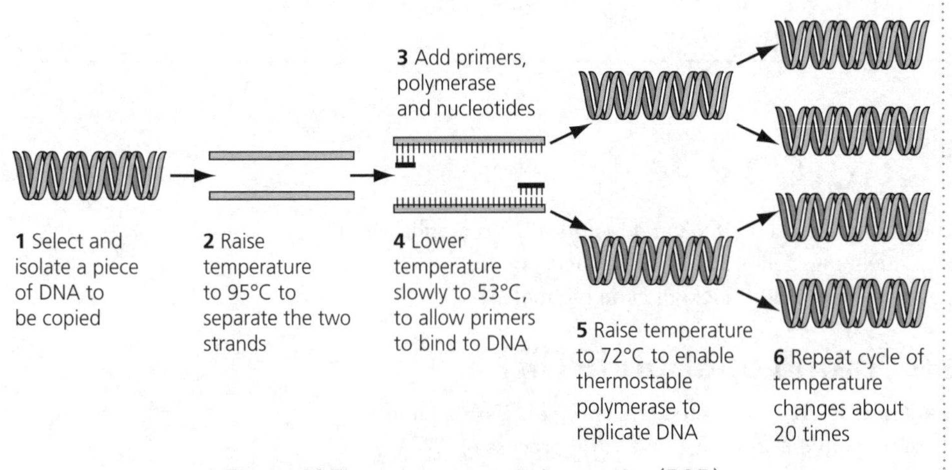

Figure 22 The polymerase chain reaction (PCR)

Use of the polymerase chain reaction

The technique is used by forensic scientists and archaeologists to study minute samples of DNA, and may be used in **genetic (DNA) fingerprinting** (also called **DNA profiling**) and in any situation where a sample of DNA needs to be amplified.

Genetic (DNA) fingerprinting

The lengths of DNA (genes) that code for polypeptides (proteins) account for only 2% of the chromosomal DNA. Further, this DNA (the exons) does not vary much. This is because differences in the nucleotide sequences (the genetic code) could result in non-functional proteins.

Knowledge check 39

The enzyme most commonly used in the PCR is **Taq DNA polymerase** extracted from the hot-spring bacterium *Thermus aquaticus*; it is not denatured at temperatures up to 80°C. Explain why it is important that the DNA polymerase used in the PCR is thermostable.

Anneal To bind complementary sequences of single-stranded DNA by hydrogen bonding, e.g. a primer or a probe is annealed to single-stranded DNA.

Knowledge check 40

Which enzyme, used in natural DNA replication, is missing in the PCR process?

Satellite repeat sequences

It is the non-coding DNA in (introns) and between genes that is used in genetic (DNA) fingerprinting. This is because it contains variable nucleotide sequences that are similar in related individuals but quite distinct in unrelated individuals. This variable non-coding DNA is known as satellite DNA and contains short DNA sequences which are repeated many times:

- Minisatellites usually contain 20–50 base pairs. The minisatellite can be repeated from 50 to several hundred times.
- Microsatellites usually contain 2–5 base pairs. The microsatellite can be repeated between 5 and 15 times.

Microsatellite repeat sequences (MRSs, also called short tandem repeats, STRs) are often used as **genetic markers** in DNA fingerprinting since the pattern of repeat sequences in any individual (other than an 'identical' twin) is unique. The same satellite 'repeat regions' occur at the same positions on both chromosomes of a homologous pair, though the number of times the sequence is repeated on each chromosome of the pair may differ.

Knowledge check 41

Identify the microsatellite sequences in the nucleotide sequence shown below:

CTAGCGATAGATAGATAGATAGGTAGTCC

Knowledge check 42

Why are satellite regions important in genetic fingerprinting?

Genetic markers
DNA sequences, with known locations on chromosomes, which are points of variation and so used to identify individuals. Examples include microsatellite repeat sequences and single nucleotide polymorphisms (SNPs) (see page 35).

DNA fingerprinting using restriction enzymes and probes

The traditional method of DNA fingerprinting uses restriction endonucleases to cut satellite repeat sequences from the DNA which are then detected using DNA probes.

Restriction endonucleases

Restriction endonucleases (also called **restriction enzymes**) cut the sugar–phosphate backbone of DNA at specific nucleotide (base) sequences. Bacteria produce these enzymes to counter attack by viruses (bacteriophages). They do this by cutting bacteriophage DNA into smaller, non-infectious fragments.

There are many different restriction enzymes produced by different species of bacteria. Each enzyme cuts DNA at a specific base sequence — the recognition (or restriction) site (see Table 4). The enzymes may make straight cuts (leaving **blunt ends**) or staggered cuts (leaving **'sticky' ends**).

Table 4 Some examples of restriction endonuclease enzymes

Restriction enzyme	Bacterial origin	Recognition site	Type of cut end
*Hae*III	*Haemophilus aegyptius*	GG¦CC CC¦GG	Blunt
*Alu*I	*Arthrobacter luteus*	AG¦CT TC¦GA	Blunt
*Eco*RI	*Escherichia coli*	G¦AATTC CTTAA¦G	'Sticky' (staggered)
*Hin*dIII	*Haemophilus influenzae*	A¦AGCTT TTCGA¦A	'Sticky'

Restriction enzymes are also used to isolate a gene for transfer into another organism (see page 39).

DNA probes

A **DNA probe** is used to locate a particular section of DNA. It consists of a short length of single-stranded DNA with a specific nucleotide (base) sequence. The probe anneals (binds) by base pairing to a complementary region of single-stranded target DNA. In order to be able to detect the probe and the target DNA to which it is attached, the probe has a fluorescent or radioactive tag. Detection takes place using ultraviolet light for fluorescent probes or an X-ray film for radioactive probes.

Knowledge check 43

Distinguish between a 'DNA primer' and a 'DNA probe'.

DNA probes are used to detect specific regions of DNA in genetic fingerprinting, defective alleles in genotype testing (genetic screening) and genetically modified cells.

Using restriction enzymes and probes, the stages are:

1 *Extracting DNA from a sample.* The DNA from a sample of blood (white blood cells), or other tissue, is extracted. In diploid eukaryotes, this will contain the DNA of two sets of chromosomes, one maternal and one paternal. If the sample of DNA is very small, more copies are made using the PCR.

2 *Using restriction enzymes to cut out repeated sequences.* Appropriate restriction enzymes are used to cut the extracted DNA into fragments at specific base sequences at sites either side of the satellite repeat sequences selected for analysis. The repeat sequences remain intact, as shown for one microsatellite in Figure 23.

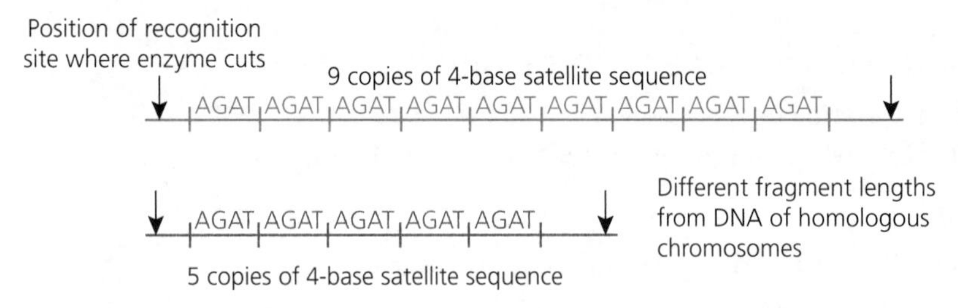

Figure 23 A restriction enzyme cuts at sites either side of a microsatellite

If there are a large number of repeats then the resultant fragments will be long; if there are few repeats, shorter fragments will be produced.

3 *Using gel electrophoresis to separate DNA fragments.* The fragments of DNA are separated by **gel electrophoresis** on the basis of size — smaller fragments move further.

4 *Transferring the DNA fragments onto a nylon sheet.* The separated DNA fragments (heat-treated to make them single-stranded) are transferred onto a nylon sheet by pressing it against the gel. This is called Southern blotting. At this stage, the DNA fragments are not visible.

5 *Attaching labelled probes.* The sheet is washed with fluorescently or radioactively labelled DNA probes. The probes are synthesised to bind (**hybridise**) to those fragments containing the complementary repeat sequence (Figure 24).

Gel electrophoresis DNA fragments are placed in a well at the negative electrode end of a gel plate; an electric current is passed through the gel and, since DNA is negatively charged, the fragments move to the positive end, with smaller fragments moving further.

Hybridise Binding of single-stranded DNA with complementary single-stranded DNA from another source.

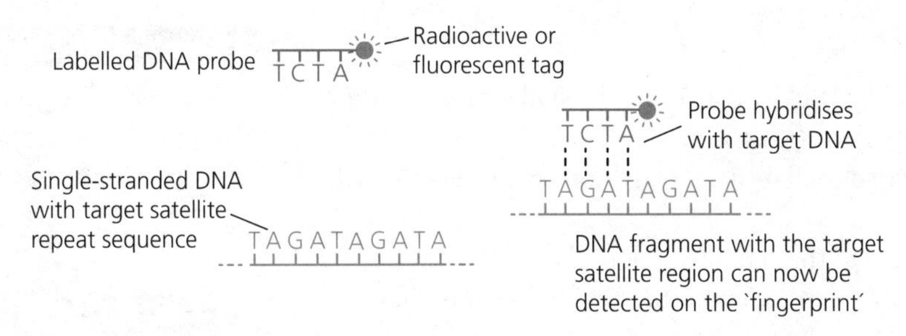

Figure 24 Use of a labelled probe to detect a DNA fragment with the target satellite region

6 *Detecting the different repeat sequences.* Only those DNA fragments with probes attached show up, using ultraviolet light for fluorescent probes or X-ray film for radioactive probes. The resulting pattern of bands is called a genetic (DNA) fingerprint and looks rather like a bar code.

The whole process is illustrated in Figure 25.

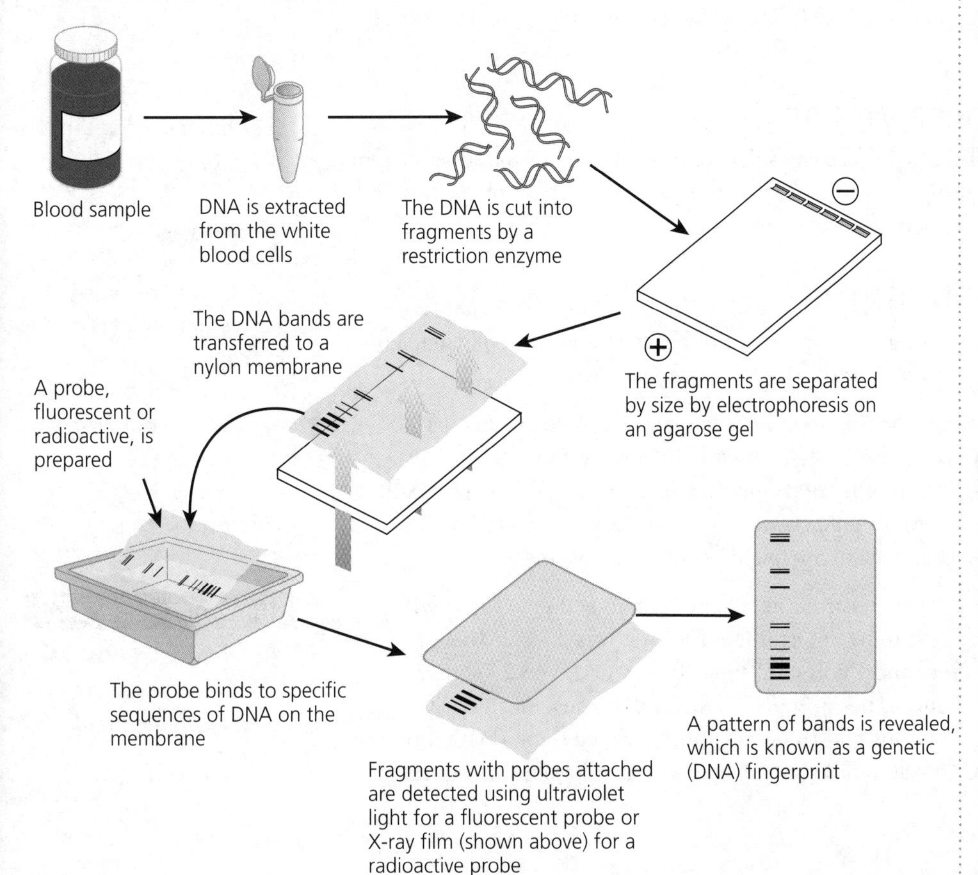

Figure 25 Stages in preparing a genetic fingerprint using restriction endonucleases and probes

Exam tip

In comparing DNA fingerprints, you can conclude that an individual's DNA is present in a sample (say at a crime scene) only if *all* the bands are included. A child's DNA will produce some bands which are found in the mother while the father must possess *all* those remaining.

DNA profiling using PCR

A more powerful technique (presently used by forensic services) makes use of the polymerase chain reaction to amplify the MRSs:

- Fluorescent DNA primers that will attach next to the region containing the MRSs are synthesised.
- With primers annealed at each side of the satellite region, the PCR is used to replicate large numbers of DNA fragments containing the MRSs. This allows minute quantities of source material to be analysed.
- The labelled fragments of DNA are separated using gel electrophoresis.
- In the UK Forensic Science Service, this is completed for ten different microsatellites, each with four or five base repeats, with an additional marker for determining gender.
- The position of the DNA fragments is revealed as a pattern of fluorescent bands due to the fluorescent tags on the DNA primers flanking the microsatellite regions. The bands are detected using a laser scanner.
- The results are displayed in a graph of fluorescence against fragment size — the DNA profile.

Knowledge check 44

Explain why forensic scientists often use the PCR when producing a genetic fingerprint.

Exam tip

Using the PCR, no part of the DNA other than the selected satellite regions is replicated. In the technique using restriction enzymes, parts of the DNA between the satellite regions will be released, though these will not show up since labelled probes will not attach to them.

Use of genetic fingerprinting

The technique is a powerful tool in forensic science, in settling paternity disputes, in establishing family relationships, and in studying the genetic diversity of species and the evolutionary relationship between taxonomic groups.

Genome sequencing

Gene sequencing involves finding the sequence of bases in a DNA molecule. This can be done for a fragment of DNA or for the entire genome.

The **genome** is the total genetic make-up of an organism. For a virus, this is the genetic information contained in its DNA strand (or RNA strand in retroviruses), while for a prokaryote (bacterium) it is the composition of the single loop of naked DNA and any plasmids present. For eukaryotic organisms, genome is defined as the complete nucleotide (base) sequence in a haploid set of chromosomes.

Exam tip

You should be aware that mitochondria and chloroplasts contain DNA and so the genome of a eukaryote may include the base sequence of the DNA in these organelles.

While the ultimate goal was the determination of the human genome, those with smaller sequences were the first to be determined (the first was a virus, then a bacterium). The **Human Genome Project** (1990–2003) worked on 24 separate chromosome sequences (22 autosomes and both X and Y chromosomes). Early sequencing methods were laborious and time-consuming. Since then, the techniques have advanced — they are now automated and sequences can be obtained quickly and more cheaply.

Knowledge check 45

Explain precisely what is meant by the term 'genome' in a diploid organism.

Exam tip

The advances in sequencing technology are startling. While the first human sequence took 13 years to complete and cost an estimated $2.7 billion, a human genome can now be sequenced in a period of weeks for less than $1000.

Genome sequencing has been undertaken for a wide range of organisms. These include:

- organisms that are used in genetic research, such as the fruit fly *Drosophila melanogaster*, used extensively in the study of heredity, and the mouse *Mus musculus*, used for much of the research into gene function (see page 46)
- organisms that can cause disease, including the bacterium *E. coli*, various strains of which are pathogenic, and parasites such as *Plasmodium*, which causes malaria
- organisms that can provide valuable information about the evolutionary ancestry of the modern human *Homo sapiens*, the chimpanzee *Pan troglodytes* (our closest living relative) and the Neanderthal *Homo neanderthalensis* (an extinct species of hominid).

Exam tip

While there are about 21 000 genes, there are many more proteins produced in human cells (due to alternative exon-splicing — see page 23). The entire set of proteins expressed by a genome is called the **proteome**.

The human genome

Since the initial project on the human genome, thousands of complete and partial genomes have been sequenced to obtain detail about human genes and the allelic variations that can exist. Findings about the human genome include the following:

- The genome consists of more than 3 billion base pairs (on a haploid set of chromosomes).
- 98% of the DNA does not code for proteins (though about 3% has a regulatory function, e.g. control of DNA replication or signals controlling when genes are expressed).
- More than half of the non-coding DNA consists of variable repeat sequences (an analysis of which forms the basis of genetic fingerprinting — see page 31).
- 2% of the DNA codes for proteins and includes an estimated 21 000 genes.
- Single nucleotide (base) substitutions (see page 26) occur in the coding and non-coding areas of the genome — these small variants are called **single nucleotide polymorphisms (SNPs)**.
- In the polypeptide-coding genes, SNPs represent allelic differences and are used as markers of genetic variation.
- SNPs vary in only about 0.1% of the human genome, though this accounts for thousands of different alleles, the genetic variants which influence how people differ.
- SNP alleles have been identified which are linked with genetic disease, genetic predisposition to disease and to different responses to drugs.

Single nucleotide polymorphisms (SNPs) Sites in the genome where the DNA sequence differs by a single base. Used as markers of genetic variation (including faulty alleles associated with genetic disease).

Exam tip

The terms base substitution, SNP and allele are related. A base substitution is a form of mutation whereby one base replaces another; SNPs are the alternative nucleotide (base) substitutions that may be found in the genome; alleles are alternative forms of a gene, often as a result of base substitution (as in sickle-cell anaemia), i.e. SNPs which influence the genetic code.

Applications of genome sequencing

Some of the applications of genome sequencing projects are summarised below.

1. *Creating phylogenetic trees.* When the genomes of different species have been sequenced, comparisons can be made to determine evolutionary relatedness and so generate more accurate phylogenetic trees. The sequences can be matched and degrees of similarity calculated. Close similarities indicate recent common ancestry.
2. *Detecting genetic disease.* Some alleles of a single gene have been linked with **genetic disease**, e.g. cystic fibrosis is caused by the recessive allele of the *CFTR* gene. Knowing the sequence of such alleles allows gene probes to be synthesised which can then be used to test for their presence in the newborn or in fetal cells

(see 'Genetic testing using microarrays' on page 36). Gene therapy (see page 45) may then be offered.

3 *Detecting genetic predisposition to certain diseases.* The genetic basis of many medical conditions is more complex, involving the alleles of many genes (polygenes — see page 59). Such genes possess alleles which are **genetic risk factors** and are associated with **genetic predisposition** towards certain diseases such as diabetes, cancer, heart disease, asthma and mental illness. For example, breast cancer is linked with alleles of two genes, *BCRA1* and *BCRA2*, which do not necessarily cause cancer but increase the risk of it happening in later life. Knowing the base sequence of such alleles provides a means of testing for their presence and assessing the risk of disease occurring.

4 *Analysing the genomes of pathogens.* Sequencing the genes of pathogens enables doctors to find out the source of an infection and to identify antibiotic-resistant strains of bacteria. It also enables scientists to track the progress of a potential epidemic, e.g. genome sequencing of the Ebola virus contributed to an understanding of its transmission during the epidemic of West Africa in 2015. And it enables pharmacologists to design drugs which more readily fit onto a pathogenic enzyme, thus disabling it and rendering the pathogen harmless (see below).

5 *Creating 3D models of protein structure.* When the base sequence of a gene has been obtained, the genetic code (Table 3) can be used to determine the primary structure (amino acid sequence) of the protein produced. Molecular modelling software, using information on the bonding between amino acids in the chain and so determining folding arrangements, then allows the secondary, tertiary and quaternary structures to be predicted. Such precise 3D images facilitate the production of therapeutic drugs, frequently enzyme inhibitors, to counter pathogens or parasites. Drugs are more precisely designed to fit into the binding pocket of an enzyme found in the pathogen, but not the host.

6 *Producing personalised medicine.* Some alleles have been uncovered which influence metabolism in such a way that individuals respond differently to toxins and drugs (including medicines). Knowing the base sequences of such alleles allows genetic testing so that drugs that are prescribed match the individual's genetic profile (see 'Pharmacogenetics' on page 38).

Genetic testing using microarrays

Genetic testing (also referred to as **genetic screening**), involving **microarrays**, commonly determines:

- genotypes (allele variants) in multiple regions of a genome. Defective alleles (frequently SNPs) have distinctive base sequences and their detection is important in diagnosing genetic disease and in determining whether an individual is a carrier of genetic disease. It is important to test simultaneously for different defective alleles, e.g. more than 100 defective alleles are linked with cystic fibrosis
- **gene expression** of a large number of genes simultaneously. Detecting only the genes which are switched on in a cell is important in the diagnosis of cancer, where certain genes (oncogenes), inactive in normal cells, become activated to cause uncontrolled cell division (a tumour). The presence of mRNA in the cytoplasm indicates that a gene is being expressed

Gene expression A gene is expressed when it is switched on so that mRNA is transcribed to encode the synthesis of a protein. Only some genes are switched on in each tissue type.

Microarrays (**gene chips** or **DNA chips**) consist of hundreds to tens of thousands of different DNA probes (see page 32) anchored within 'spots' to a solid surface. Each spot of the array contains many molecules of a particular probe, while different spots possess probes complementary to different known genes (Figure 26).

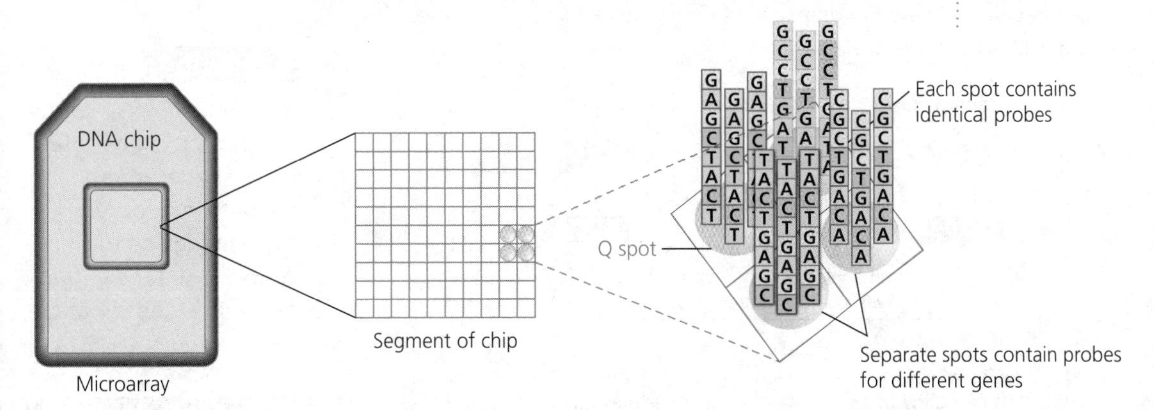

Figure 26 A microarray

The principle behind microarray technology is hybridisation between the DNA probe and any complementary single-stranded DNA obtained from the tissue tested. In genotyping, DNA from a person being tested is extracted, and genes cut out (or copied), made single-stranded and labelled with a fluorescent dye (or chemiluminescent tag). If, after adding to the microarray, a DNA fragment containing a gene hybridises with a probe, a colour change occurs (or light is emitted if the tag is chemiluminescent). Computer analysis of the results will reveal a whole range of genes in a person's genome, including any faulty alleles of particular genes.

Knowledge check 46

What gene base sequence would hybridise with the probes in spot Q in Figure 26?

Exam tip

To determine whether an individual possesses a mutation for a particular disease, a sample of the patient's DNA may be compared to a control sample — one that does not contain a mutation (in the gene of interest).

If testing for gene expression, the mRNA is extracted from the tissue and converted, using reverse transcriptase enzyme (page 40), into single-stranded copy DNA (cDNA) molecules, each with a fluorescent tag. The labelled cDNA molecules are added to the microarray and will hybridise with any complementary probe. Hybridisation of cDNA and probe indicates that the gene was expressed (active) in the tissue.

Exam tip

Note that it is the DNA obtained from the tissue that is labelled (not the probe, as is the case in genetic fingerprinting).

Microarray gene expression is used to diagnose the activity of oncogenes in different types of cancer (see Figure 27). Labelled cDNA is produced from mRNA extracted from healthy and from cancer tissue, with different fluorescent dyes used to tag the cDNA from each tissue, say 'healthy' green and 'cancer' red. A mixture of both sources of cDNA is added to a microarray so that both 'healthy' and 'cancer' cDNA may hybridise to the probes. The microarray is scanned with a laser, exciting the fluorescent tags on the cDNA and generating signals, the strength of which depends on the amount of cDNA binding to the probes on each spot. The signals, indicating the expression level of various genes, are analysed: spots with both green and red tags signalled (showing yellow in Figure 27) are of little interest since they represent genes which are active in both the healthy and cancer tissue; spots signalling green represent genes which are active in the healthy tissue only and are switched off in cancer tissue; while spots signalling red represent genes which are switched on in cancer tissue. Some of the genes which are switched on in the cancer tissue may be oncogenes (epigenetically activated perhaps), which may be deactivating some genes (those green) and activating others.

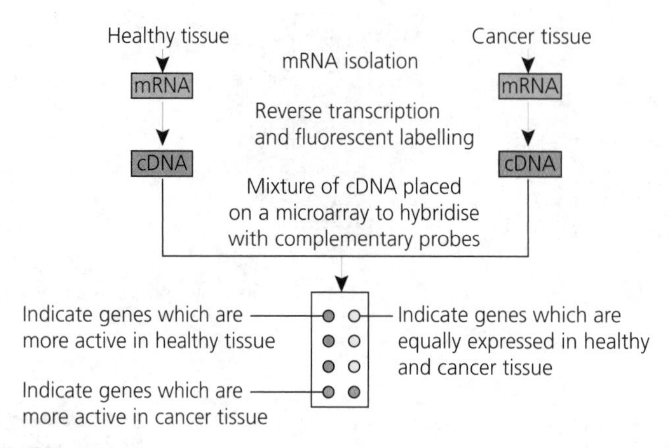

Figure 27 Using a microarray to test for gene expression in cancer tissue

Pharmacogenetics

It has long been known that many drugs work in only a proportion of the people taking them, and may produce adverse effects in some people. This is often due to large individual variations in the rate of metabolism of drugs. Drugs are metabolised in the body by enzymes, some of which inactivate drugs, though others make an inactive drug more active. These enzymes are encoded by genes that have many allelic (SNP) variants. **Pharmacogenetics** is the study of genetic differences in drug metabolic pathways, which can affect individual responses to drugs.

Inactivation of drugs:

CYP2 enzymes are responsible for the breakdown of many drugs, though their activity depends on which of a series of alleles is responsible for their production — some alleles encode for rapid-metabolising enzymes while others produce poor metabolisers. The clinical effect of a given dose of drug is greater in a poor metaboliser and much less in a rapid metaboliser because of the different rates of elimination of the drug. If a standard dose is used, rapid metabolisers may get no benefit from the drug, while poor metabolisers may suffer effects of an overdose.

Activation of drugs:

Codeine, a drug prescribed for pain relief, must first be converted by an enzyme (CYP2D6) to its active form morphine before becoming effective. The gene encoding for this enzyme is *CYP2D6*, variant alleles of which are classified on the basis of enzyme activities, i.e. one allele may produce an enzyme that metabolises codeine slowly, while another may encode a highly efficient enzyme. A poor metaboliser will get no pain relief from a standard dose of codeine, while rapid metabolisers are at increased risk of adverse effects (such as breathing problems).

Exam tip

Read questions carefully. It is possible for the experimental cDNA to incorporate a green tag and the control a red tag — the opposite of the situation in Figure 27.

Knowledge check 47

Explain how allelic differences would affect the activity of an enzyme.

Exam tip

Notice that, depending on the medical drug concerned, an enzyme may be an activator or an inactivator.

Knowledge check 48

A patient possesses alleles that encode for a highly active form of the CYP2D6 enzyme. Should they be prescribed a higher or lower dose of codeine? Explain your answer.

Exam tip

Letters representing a gene are italicised (*CYP2D6*), while when representing an enzyme they are not (CYP2D6).

An analysis of the genetic profile of an individual patient can be undertaken using gene probes and microarrays. This knowledge would predetermine which drug (and which dosage) provides the best chance of achieving the desired therapeutic effect while reducing the likelihood of adverse effects.

This is used in **personalised medicine** to decide which drug should be prescribed when several different ones are available. Most progress in personalised medicines has been made with drugs for treating cancers. Approximately 30% of breast cancers involve a mutation of the *HER2* gene (such individuals are said to be *HER2*-positive). The activity of this gene can be shut down by certain drugs, e.g. trastuzumab (marketed as Herceptin). A genetic test of excised breast tissue is used to determine the presence of the mutation to ascertain whether trastuzumab is an appropriate treatment. It is important that trastuzumab is restricted to *HER2*-positive individuals as it is expensive and has been associated with cardiac toxicity. Drugs such as trastuzumab, since they are created to match an individual's genetic profile, are sometimes called '**designer drugs**'.

Personalised medicine Choosing treatments for people based on knowledge of individual personal genetic profiles rather than using the same treatment for all.

In future, with DNA sequencing occurring relatively rapidly and cheaply, it should become increasingly common for clinicians to analyse the genome of their patients (and the genome of any invading pathogens).

Genetically modified organisms

Genetically modified organisms (**GMOs**) are also called **genetically engineered organisms**. Their production involves **gene transfer**, which is the transfer of a gene from a donor organism to a recipient organism. Since gene transfer involves combining DNA from different sources, it is referred to as **recombinant DNA technology**.

Genetically engineered microorganisms

There are a number of stages in manipulating genes to produce genetically modified organisms and, specifically, **genetically engineered microorganisms** (**GEMs**): obtaining the required gene; inserting the gene into a vector; inserting the vector into a host cell; and identification of the host cells that have taken up the gene.

Obtaining the required gene

There are two main methods used to obtain a gene:

1 Using restriction endonuclease to cut the gene out of chromosomal DNA (Figure 28):
   - The chromosomal DNA is cut into fragments using an appropriate restriction endonuclease enzyme, which cuts either side of the gene (see page 31).
   - The DNA fragment containing the required gene is identified using a complementary gene probe (see page 32).
   - A restriction endonuclease that cuts in a staggered fashion to produce 'sticky ends' is most useful. If the vector is opened with the same restriction enzyme, then the exposed bases of both are complementary and so will more readily attach through base pairing.

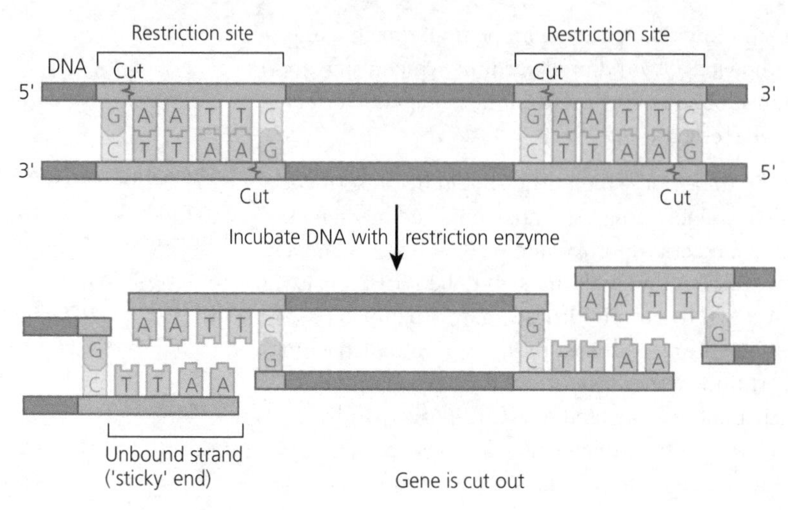

Figure 28 Use of restriction endonuclease enzymes to cut the gene out of chromosomal DNA

Knowledge check 49

Different restriction enzymes have different recognition (restriction) sites. Explain this.

Knowledge check 50

Explain why a single type of restriction endonuclease enzyme cannot be used to cut out all the genes on a chromosome.

Knowledge check 51

A gene (DNA) probe contains the base sequence ATAGCAGTCG. (a) What can you deduce about the base sequence of the target DNA? (b) How will the target DNA, with its attached probe, be identified?

2 Using reverse transcriptase to produce DNA from mRNA (Figure 29):
   - Messenger RNA is obtained from cells where the gene concerned is actively synthesising protein (such cells possess many copies of the mRNA).
   - The enzyme reverse transcriptase uses the mRNA as a template to produce a complementary single strand of DNA (complementary or **cDNA**) from free DNA nucleotides.
   - Double-stranded DNA is made from the cDNA using the enzyme DNA polymerase. Only if the DNA is double-stranded can it be annealed to the double-stranded plasmid vector.

Synoptic links

Bacteria and viruses

In AS Unit 1 you learned about the structure of bacteria and viruses, and the role of reverse transcriptase in retroviruses.

Exam tip

Human genes synthesised from mRNA will not contain introns and so will be able to direct protein synthesis when inserted into bacteria. This makes it a highly advantageous method. Genes obtained using restriction endonuclease enzymes will need to have their introns removed before being inserted into bacteria.

Knowledge check 52

Suggest the advantage of manufacturing a gene from mRNA rather than cutting it out of the DNA genome.

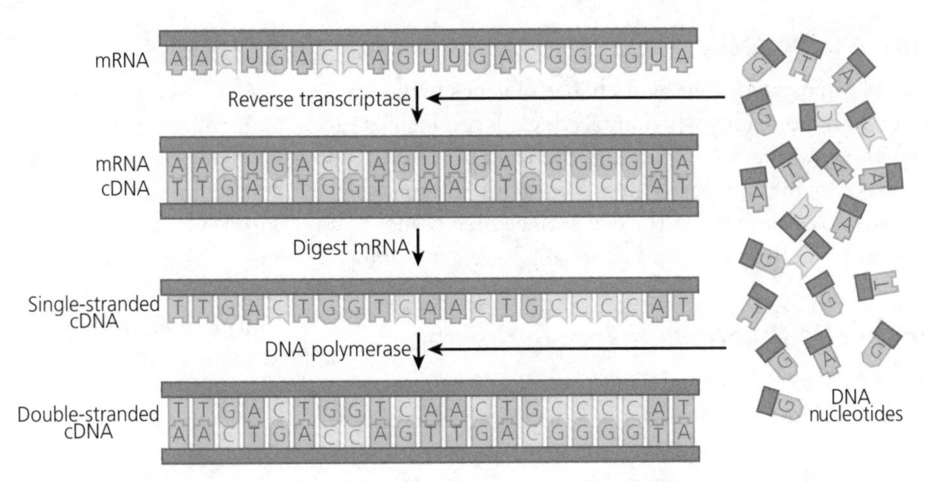

Figure 29 Use of reverse transcriptase to produce DNA from messenger RNA

Inserting the gene into a vector

Once the gene has been obtained it is inserted into a **vector**. The main types of vector are bacterial **plasmids** and viruses, though there are others.

Vector A delivery tool to carry the gene into the host cell where it can be replicated and expressed.

Plasmid A small, circular piece of extrachromosomal DNA (i.e. found outside the main loop of DNA) that occurs in certain bacteria (and yeast). Plasmids carry extra genes — R-plasmids carry genes that provide resistance to naturally occurring antibiotics.

1 **Bacterial plasmids** are the most common vector for inserting genes into bacterial cells. The method of inserting DNA containing the required gene into a plasmid involves the following processes:
   - The plasmid is cut open using the *same restriction enzyme* used to cut the DNA fragment out of the donor DNA. The sticky ends of the two types of DNA contain complementary base sequences (if the gene was synthesised from mRNA, sticky ends are added).
   - The plasmid DNA and gene DNA anneal. Hydrogen bonds form readily between the complementary bases of the sticky ends and **DNA ligase** catalyses the formation of phosphodiester bonds (covalent bonds) between the sugar–phosphate backbones of the plasmid DNA and the gene DNA. The gene is said to be **spliced** into the plasmid.

   Any DNA that has 'foreign DNA' inserted into it is called **recombinant DNA**, so the plasmid is now a **recombinant plasmid**.

   This process is shown in Figure 30.

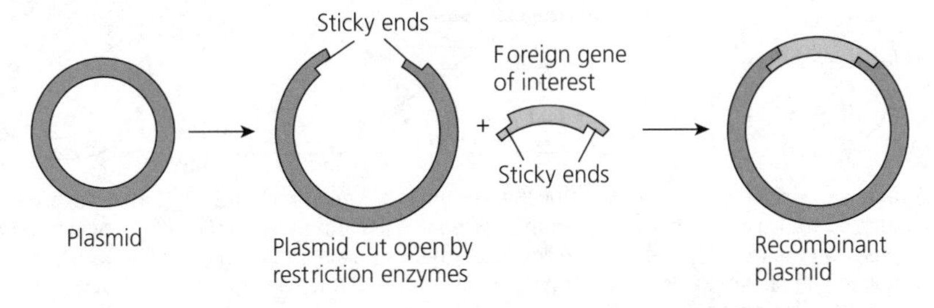

Figure 30 Transferring a gene into a plasmid

2 **Viruses** are adapted to insert their genetic material into a host cell — for example, a bacteriophage 'injects' its DNA into a bacterium. A bacteriophage that has a DNA fragment spliced into its DNA will transfer that recombinant DNA into the bacterial cell.

Inserting the vector into a host cell

Bacterial cells naturally take up plasmids (allowing the exchange of genes between bacterial cells). However, they do so more readily when induced. Cells are incubated with *calcium ions* and treated with a *heat shock* (temperature rise from 0°C to 42°C), which makes the cell-surface membrane permeable to plasmids; alternatively, a short electrical pulse may be used to open pores in the cell-surface membrane (electroporation). Bacteriophages are an effective way of delivering large genes into bacterial cells.

Identification of the host cells that have taken up the gene

Only a few of the bacterial cells will take up a recombinant plasmid — the rate of take-up may be as low as 1 in 10 000. Most will either fail to take up a plasmid at all or will take up an original, non-transformed plasmid. Bacteria that have taken up the recombinant plasmids are called **transformed bacteria**. Identifying successful take-up involves using a second, separate gene (**marker gene**).

1 **Using antibiotic-resistant marker genes.** Some plasmids (R-plasmids) carry genes that confer antibiotic resistance to the bacteria. One particular R-plasmid has genes for resistance to two antibiotics, ampicillin and tetracycline. These may be used as marker genes. For example, an appropriate restriction enzyme cuts in the middle of the tetracycline resistance gene and the 'desired' gene is inserted. The transformed plasmid now contains an active ampicillin resistance gene but an inactive tetracycline resistance gene (Figure 31). Bacteria are detected according to the following criteria:
   - Those that failed to take up plasmids are sensitive to both ampicillin and tetracycline.
   - Those that take up the original plasmids are resistant to both ampicillin and tetracycline.
   - Those that take up the recombinant plasmids are resistant to ampicillin but not to tetracycline.

Exam tip

It is important to isolate and culture only the transformed bacteria for further culture since non-transformed bacteria are of no commercial value and compete with the transformed bacteria for nutrients.

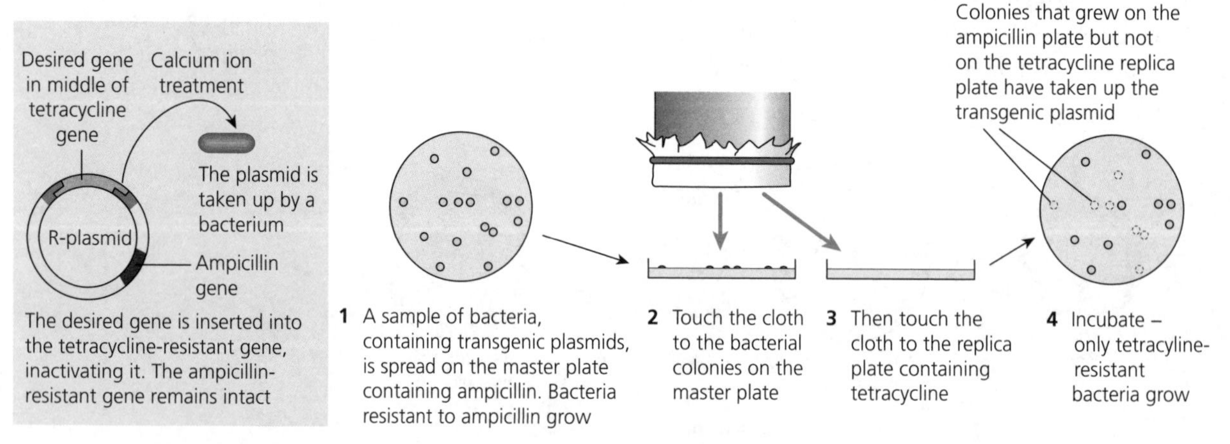

Figure 31 Identifying transformed bacteria using antibiotic-resistant marker genes and replica-plating

The bacteria are cultured on agar plates. Each bacterium multiplies to form a colony. The colonies on the agar plate are then replica-plated onto a plate containing ampicillin and a plate containing tetracycline. Replica-plating involves 'blotting' the original plate with a pad and then pressing this against the surface of a fresh plate so that a few cells from each colony are transferred. The bacteria that survive on the ampicillin plate *only* are the transformed bacteria.

2 **Using fluorescent marker genes.** This is a more recent method of finding out whether bacteria have taken up the desired gene. Some organisms (e.g. jellyfish) possess genes that produce fluorescent proteins. Such a gene is isolated and attached to the gene of interest — this hybrid is marked since it produces fluorescent protein. The attached genes are inserted into the vector (plasmid) and carried into the bacterial cell. Take-up of the desired gene is readily identified as transformed bacteria will glow under ultraviolet light.

Those bacteria identified as being transformed are then transferred to a sterile nutrient medium and allowed to increase in number. This is called **gene cloning** and increases the productivity of the desired product, which is then extracted and purified.

Knowledge check 53

An R-plasmid, containing genes for resistance to the antibiotics chloramphenicol and kanamycin, was cut open with an appropriate restriction enzyme and the gene of interest was inserted in the chloramphenicol-resistant gene. What would be the criterion for subsequently selecting the transformed bacteria?

Genetically modified plants

Foreign genes are inserted into plant cells using a range of methods.

- The most common method of transferring genes into plant cells is to use the common soil bacterium *Agrobacterium tumefaciens*, which readily invades damaged plant tissue and causes tumour-like growths. On entering the damaged tissue the bacterium's **tumour-inducing (Ti) plasmid** is transferred into a plant cell. For gene transfer, the desired gene is spliced into the plasmid, which is readily taken up by the plant cell provided that its cellulose wall has been removed by treatment with the enzyme cellulase.
- Minute pellets that are covered with DNA carrying the desired gene are shot through the cellulose walls into plant cells using a particle gun (**gene gun**).
- **Plant viruses** are sometimes used. They infect cells by inserting their nucleic acid and therefore new genes can be transferred into the cell.

The DNA fragment containing the foreign gene is incorporated into the plant genome. The new gene is expressed using the plant cell's synthetic machinery (ribosomes, mRNA etc.) — the plant has been genetically modified.

A gene for antibiotic or herbicide resistance, or a fluorescent marker gene, is inserted along with the desired gene. In the former case, treatment with the antibiotic will kill the non-transformed cells, leaving only those that have been transformed. With a fluorescent marker gene, transformed cells will show up under ultraviolet light. Transgenic plant cells are then tissue-cultured and entire genetically modified plants are grown from this tissue.

Knowledge check 54

What type of marker gene could be used for the identification and selection of genetically modified plant cells?

Genetically modified animals

Several techniques can be used to artificially introduce genes into animal cells:

- A common method of introducing DNA into an animal cell is by **electroporation**, a technique in which the cell membrane is temporarily disrupted through treatment with a high-voltage shock.
- Another approach is to coat DNA in minute artificial lipid vesicles, called **liposomes** (Figure 32), which may adhere to the cell-surface membrane and pass the DNA into the cell in a similar way to endocytosis.
- **Viruses** (adenoviruses and retroviruses) can be used to insert genes into animal cells. Adenoviruses are human viruses that can cause respiratory diseases. Therefore, they have to be altered genetically so that the host cells are not destroyed. Adenoviruses are particularly useful for delivering genes to patients in gene therapy. Retroviruses have RNA as their genetic material. When their RNA is delivered into a host cell it is copied to DNA and the DNA is incorporated into the host's chromosome.
- It is possible to inject DNA directly into the nucleus of a fertilised egg. This technique is called **microinjection**.

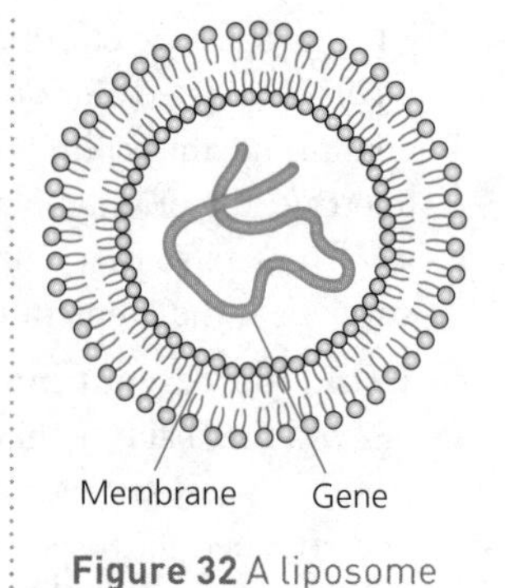

Figure 32 A liposome

Exam tip

Be aware of different terms for essentially the same thing. A genetically modified organism may also be called a genetically engineered organism or a **transgenic organism**.

Uses of genetically modified organisms

There are a huge number of GMOs that have been engineered for use in medicine, agriculture and industry. Some examples are shown in Table 5.

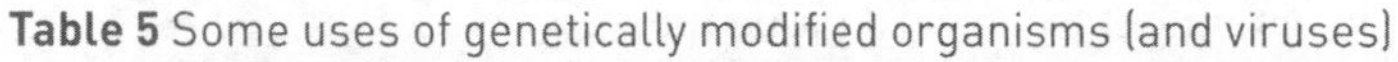

Table 5 Some uses of genetically modified organisms (and viruses)

Genetically modified organism	Transferred gene(s)	Purpose
Herpes (cold sore) virus	Gene encoding a protein (GM-CSF) that stimulates the immune system	The engineered herpes virus (altered so incapable of causing cold sores), called T-VEC, attacks skin cancer cells
Bacteriophage	Gene encoding for a bacterial protein (lexA3) impairs the bacterium's ability to repair damaged DNA	The modified phage, M13, makes infectious bacteria more vulnerable to DNA-damaging drugs and to antibiotics (to which they might otherwise be resistant)
Bacteria (*Escherichia coli*)	Human gene (produced from mRNA isolated from pancreatic tissue) for the production of the hormone insulin	Insulin is produced for use by people who cannot produce their own (suffer from type I diabetes mellitus)
Bacteria (*E. coli*) and fungi (*Aspergillus niger*)	Cow gene for the production of the enzyme chymosin	Chymosin is a coagulating enzyme used in the commercial production of hard cheese; most is genetically engineered
Yeast (*Saccharomyces cerevisiae*)	Human gene coding for the production of the protein alpha-1-antitrypsin	Alpha-1-antitrypsin is used to treat patients with hereditary emphysema
Soya (*Glycine max*)	1 A gene that makes the plants resistant to a specific herbicide (e.g. glyphosate) 2 Another gene that confers tolerance to drought and soil salinity	1 Spraying with the herbicide kills competing weeds and so the productivity of the resistant crop plant is increased 2 Tolerance to drought etc. allows plants to grow in unfavourable conditions

→

Table 5 *continued*

Genetically modified organism	Transferred gene(s)	Purpose
Maize (*Zea mays*)	The *Bt* gene from the bacterium *Bacillus thuringiensis*	The gene produces a protein that is toxic to some insects (mainly caterpillars), so reducing the use of insecticides
Cows	Human gene *LALBA* that codes for the synthesis of the protein alpha-lactalbumin	Produces human protein-enriched milk, more nutritionally balanced than natural cow's milk (for babies or the elderly with special nutritional needs)
Sheep (ewes)	Human gene for the blood-clotting protein Factor VIII	Factor VIII is extracted from the milk of transgenic ewes and used as a treatment for haemophilia A

Knowledge check 55

What 'tools' would a genetic engineer use for (a) cutting a fragment from a DNA molecule, (b) identifying a DNA fragment containing the gene of interest, (c) joining together different lengths of DNA, and (d) producing DNA from mRNA?

Genetically modified mice, for use as **model organisms** in research on gene function, are discussed on page 46.

Exam tip

You should be familiar with these examples of gene transfer, though expect any question on genetic engineering to be unfamiliar.

Model organisms Species extensively studied to understand particular processes with the expectation that discoveries provide insight into the workings of other organisms.

Gene therapy

Many genetic diseases are caused by defective alleles which are recessive. Gene therapy is the treatment of such diseases by introducing the functional allele of the gene into the affected cells.

Exam tip

You could be asked to outline how a defective allele can be identified. The process is as follows: a DNA sample is obtained from the individual; a gene probe is used to detect the mutated base sequence of the defective allele in the individual's genome; if the mutated sequence is present, the probe will bind to it and flag the defective allele.

Knowledge check 56

Explain why gene therapy could not work with a defective allele that is dominant, such as Huntington's disease.

Examples of gene therapy

Haemophilia B is the result of a faulty allele of a gene found on the X chromosome. As a result, factor IX, a protein required for blood clotting, is not produced. Instead of intravenous injections of expensive clotting factor every 2–3 days, gene therapy attempts to provide a single treatment with the functional gene introduced into liver cells via a vector. Separate trials have used, with some success, different viral vectors — adeno-associated-virus and lentivirus (a retrovirus). The vector, carrying the gene, is injected directly into the body and targets liver cells where the gene is expressed.

Cystic fibrosis is caused by alleles that fail to produce a trans-membrane conductance regulator (CFTR) protein. As a result, a thick mucus collects in the respiratory tract, interfering with gas exchange. (The pancreatic duct is also blocked, interfering with digestion in the small intestine.) The traditional treatment involved regular physiotherapy to remove mucus. In gene therapy, both adenoviruses and liposomes (see Figure 32) are used as vectors for the functional *CFTR* gene. The vector is administered as an aerosol into the patient's respiratory passages using an inhaler and enters the epithelial cells.

Knowledge check 57

Explain how liposomes combine with the cell-surface membrane of the lung epithelial cells.

Benefits and problems of gene therapy

Clinical trials have taken place with gene therapies for a number of genetic diseases. While undoubtedly progress has been made, there have also been setbacks. Perhaps most success has been with the treatment of haemophilia B where patients, in different trials, have been able to live free of the need for regular injections of clotting factor. Still, gene therapy has huge potential for the elimination of disease in which some symptoms may be treated but for which there is no cure.

Exam tip

There are two types of gene therapy. In **somatic gene therapy**, functional therapeutic genes are transferred into the body cells of a patient (e.g. with cystic fibrosis) — they will not be inherited. In **germ-line gene therapy**, functional genes are introduced into germ cells (eggs or sperm) and will be inherited by future generations, giving rise to ethical concerns.

Problems have arisen through medical side effects. Some viruses used as vectors, while modified in an attempt to disable any pathogenic effect, have been associated with disease on occasion (e.g. adenoviruses have been linked with lung infection). Viral particles and viral-modified cells may be attacked by the immune system, necessitating the use of immunosuppressant drugs.

There are also practical problems. Many cells fail to take up the new gene or, even if they do, fail to express it. Gene therapy treatments frequently need to be repeated. For example, with cystic fibrosis, treatment can be short-lived since epithelial cells of the respiratory passages do not live for long, so treatment needs to be repeated every few weeks. Also with cystic fibrosis, only cells in the respiratory passages can be treated, not cells in the pancreas.

Gene knockout (knockin) technology

A **gene knockout** is a genetically engineered organism that carries a gene that has been made inoperative, leading to a loss of gene function. One form of **gene knockin** exchanges a mutated DNA sequence for a 'normal' sequence in the gene, putting in place the function of the mutant gene. Both techniques allow the function of genes to be studied. Comparing how the knockout or knockin organism differs from individuals in which that particular gene has not been made inoperative or replaced provides information about what the gene does. The role of the gene in protein production and its metabolic and physiological influence can be determined. Both technologies can also be used in the development and testing of therapeutic drugs.

Exam tip

Any question about gene knockouts will be novel. Read the question carefully and use the information provided. You are being asked to apply your understanding of the principles of gene action.

In another form of gene knockin, a human gene is added in an organism's genome. This results in a human protein being produced in the organism.

Knockout/knockin mice

Mice are the laboratory species most closely related to humans in which the knockout technique can be performed easily. **Knockout mice** and **knockin mice** have been produced for studying human genetic diseases (such as cystic fibrosis) and also for studying genes that become defective through mutation and cause cancer. For example, it has been found that knocking out a gene for a protein called p53 greatly increases the risk that the mouse will develop cancer, indicating that this gene is involved in the control of the mitotic cell cycle.

Exam tip

Knockout and knockin mice models have been produced for most genes in the human genome.

Knockin mice have been produced with the human immunoglobulin gene introduced, enabling the mice to produce therapeutically useful humanised antibodies.

Issues surrounding the use of gene technology

Some people find gene technology controversial. In such circumstances there is a need to be as fully informed as possible in order to evaluate and discuss the issues rationally. There are obvious potential benefits and, therefore, arguments in favour of its use. However, there are potential risks that may be cited in arguments against its use. Some examples are shown in Table 6.

Table 6 Examples of the benefits and risks associated with the use of gene technology

	Potential benefits: arguments for the use of gene technology	Potential risks/concerns: arguments against the use of gene technology
Genetically engineered microorganisms (GEMs) developed to produce protein	More economic and wider production of medically important proteins, e.g. insulin	Some of the microorganisms (e.g. *E. coli*) live normally in the human gut; GEMs could escape from the laboratory and create a new strain of 'superbug'
Genetically modified (GM) plants including GM crops	Cheaper food for richer countries Possible reduction in the use of pesticides Reduction of food shortages in poorer countries	Risk of 'genetic pollution' with the spread of new genes from the modified crop to wild species, e.g. the formation of 'super weeds' Ecological concern that genetically modified plants may out-compete wild plants Concerns about allergic reactions following the consumption of GM foods
Genetically modified animals for food	Increased productivity of animals such as fish and cattle by transferring the gene for growth hormone into their genome	Concern that foreign protein, produced by transferred genes, may act as antigens (allergens) and increase the likelihood of allergies
Gene therapy	Effective treatment of genetic diseases (e.g. cystic fibrosis), relieving suffering and increasing life expectancy	Introducing genes into the human genome may disrupt the functioning of other genes, as in the appearance of leukaemia in patients treated for severe combined immunodeficiency (SCID)
Human genome research	Facilitates biomedical research	Concern that information from genome research might be used to produce 'designer babies' (e.g. for 'looks' and high IQ) Concern that an individual's genomic information (e.g. regarding susceptibility to heart disease) might become available to insurance companies
Genetic screening	People will have a better understanding of the risk of passing on a genetic disorder The fetus may be tested for the disorder before birth (called prenatal diagnostic testing)	Increased risk of stress resulting from the knowledge of being a carrier or of developing a disorder later in life (e.g. Huntington's disease) Termination of a pregnancy may not be acceptable

→

Knowledge check 58

Genes that confer resistance to herbicides or insecticides are sometimes transferred to chloroplasts rather than to the nuclei of plant cells. Suggest why this would reduce horizontal transfer of these genes into wild populations by pollen.

Table 6 *continued*

	Potential benefits: arguments for the use of gene technology	Potential risks/concerns: arguments against the use of gene technology
Gene knockout technology	Better understanding of how genes function; these genes might be implicated in a genetic disorder or might mutate to cause cancer	Large numbers of mice are used in biomedical research, many of which may be in pain; there is the view that animals have rights and that it is unacceptable to use them in this way

In order to reduce risks from the use of genetically modified microorganisms a number of **safety precautions** have been devised:

- Use of bacterial strains ill-adapted to human physiology, for example:
  - strains that grow more slowly than normal wild-type intestinal bacteria so that they are out-competed and eliminated
  - strains with a minimum temperature tolerance above human body temperature so that they will not multiply in the human body.
- Use of strains that contain 'suicide genes' which are activated if conditions move outside certain pH or temperature limits.
- Use of containment mechanisms — for example, highly efficient air filters along with regular monitoring of the atmosphere in purpose-built laboratories.

There are also **ethical concerns** over tampering with DNA of different species in ways that could never happen in nature. Some people are sceptical of the ownership of this powerful technology by a handful of multinational corporations that may be more interested in profits than in the long-term welfare of humans and the environment. There are also ethical concerns surrounding the use of genetic screening, i.e. testing for genetic disorders in parents and embryos. Germ-line gene therapy — the transfer of genes into gametes to correct a genetic disorder — is a particularly contentious issue and raises the possibility of engineering 'designer babies'.

Exam tip

Some questions raise ethical issues. This is to emphasise that good scientific practice should consider not only what we 'can do' but also whether we 'should do' it. In any discussion on ethical issues you should be able to present a rational and balanced account, with arguments both for and against.

With so many views, there is a need for the government to make decisions, i.e. there is a need for **legislation**. In the UK, the use of gene technologies is regulated strictly and research in the area of germ-line gene therapy in humans is banned.

Practical work

Gel electrophoresis of DNA and/or extraction of DNA.

Knowledge check 59

There are fears that new strains of bacteria (e.g. with antibiotic resistance) could escape from laboratories and cause outbreaks of disease. As a consequence, a variety of safety precautions has been developed. State two of these precautions.

Summary

- The polymerase chain reaction (PCR) uses a thermostable DNA polymerase in a thermal cycler to amplify minute samples of DNA.
- Genetic (DNA) fingerprinting is used to compare DNA samples, with a variety of applications that includes forensic analysis. The traditional method uses restriction endonucleases to cut DNA at specific recognition sites, and DNA probes to identify DNA fragments containing selected satellite regions.
- Genome sequencing has been undertaken for a variety of organisms. Sequencing of the human genome indicates that there are about 21 000 polypeptide-coding genes with non-coding regions (the bulk of the DNA) containing satellite repeat sequences.
- Microarrays, containing a multitude of DNA probes (each one complementary for a specific gene), are used to test an individual's genotype or the level of gene expression in an individual.
- Pharmacogenetics studies genetic differences (influencing the potency of encoded enzymes) involved in drug metabolic pathways which affect individual responses to drugs.
- The process of engineering genetically modified (transgenic) organisms involves obtaining the desired gene, either by cutting out the appropriate DNA fragment (using a restriction endonuclease) or synthesising it from isolated mRNA (using reverse transcriptase); using a vector (e.g. plasmid, virus, microprojectile, liposome) to deliver the gene into the host cell; using marker genes to identify the transformed cells.
- Bacteria are genetically engineered to produce useful proteins, for example insulin. Plants and animals can be genetically modified either to improve their productivity or for medical use. For example, plant crops can be made pest-resistant and cows can be modified to produce human growth hormone in their milk.
- Gene therapy involves treating human genetic disorders (e.g. haemophilia B, cystic fibrosis) by adding healthy genes.
- Gene knockouts provide information concerning gene function, for example that cystic fibrosis involves a transmembranal protein.
- Gene technology raises many ethical and social issues, and public awareness and debate are very important.

Genes and patterns of inheritance

A **gene** is a length of DNA that codes for a particular trait. The position of a gene on the chromosome is called the **genetic locus** (plural: loci). For each gene there are alternative forms called **alleles**.

In most plants and animals, each body cell contains two sets of chromosomes. The chromosomes exist as homologous pairs. Homologous chromosomes have the same genetic loci, i.e. they possess alleles of the same genes (one from the mother and the other from the father). If the alleles on the homologous chromosomes are the same then the individual is **homozygous**; if the alleles are different the individual is **heterozygous**.

The two alleles for a particular trait represent the **genotype**. The actual appearance of the trait represents the **phenotype**. The phenotype is determined by the genotype but may also be modified by the alleles of other genes and by **environmental factors**.

Synoptic links

Genes, chromosomes and meiosis

In AS Unit 1 you learned about the relationship between genes, chromosomes and ploidy; and the meiotic events that explain Mendel's laws of inheritance.

How the phenotype is expressed depends on which allele is dominant. A **dominant** allele has its instruction followed and so its effect is produced in the heterozygous condition. The allele that does not have its instruction followed in the heterozygous condition is said to be **recessive**.

For example, humans possess a gene that determines the ability to taste phenylthiocarbamide (PTC), a chemical that may, or may not, taste bitter. The PTC gene, *TAS2R38*, was discovered in 2003 as a consequence of work carried out during the Human Genome Project. There are two alleles: one (designated by the symbol **T**) is the tasting allele, while the other (designated **t**) is the non-tasting allele. The tasting allele (**T**) codes for a bitter taste receptor protein to which PTC can bind. The non-tasting allele (**t**) codes for a non-functional protein. **T** is dominant over **t**, since a heterozygote, **Tt**, possesses the allele **T** and produces the taste receptor protein. Things are never so simple, and environmental factors can affect PTC tasting ability — for example, having a dry mouth may make it more difficult to taste PTC and what is eaten or drunk beforehand may also affect tasting ability.

Exam tip

Many students fail to understand the distinction between 'gene' and 'allele'. A gene is a length of DNA with the base sequence that codes for a particular polypeptide (e.g. β-chain of haemoglobin). The base sequence of alleles differs in at least one base pair, resulting in polypeptides with slightly different amino acid sequences (e.g. 'normal' and sickle-cell chains).

Heredity of alleles at one locus: monohybrid inheritance

Heredity is the transfer of genetic factors from one generation to the next, i.e. from parents to their offspring. **Monohybrid inheritance** involves the inheritance of the alleles of a single gene.

Knowledge check 60

If a diploid organism has two different alleles for the same gene, is it homozygous or heterozygous?

In sexual reproduction, new individuals develop from a zygote produced by the fusion of male and female gametes. Since gametes are haploid (possess only one set of chromosomes), they contain only one allele of each gene. Thus, while a person has a pair of alleles for any genetic condition in body cells, only one allele of the pair is passed on via any one gamete. If an individual is homozygous (for example, **TT** or **tt** in PTC tasting), there can be only one type of gamete produced. If the individual is **TT** then all the gametes contain the allele **T**; if **tt** then all the gametes contain the allele **t**. If an individual is heterozygous (for example, **Tt**), half the gametes produced will contain one allele (**T**) and half will contain the other (**t**). This is the essence of **Mendel's first law**, the law of segregation of factors, which states that '*when any individual produces gametes, the alleles separate, so that each gamete receives only one allele*'. This is explained by the separation of homologous chromosomes, carrying the alleles, during anaphase I of meiosis.

When gametes combine at fertilisation to form a zygote, the alleles are restored to a pair: one from the female parent (e.g. in an egg) and one from the male parent (e.g. in a sperm cell). Fertilisation is a random event. This means that any male gamete may fertilise a female gamete. The probability of alleles combining at fertilisation depends on the frequency of the gametes containing those alleles. For example, if half the male gametes contain **T** (because the male is heterozygous, **Tt**) and half the female gametes contain **T** (because the female is heterozygous, **Tt**), then the probability of a **TT** combination in a zygote is ½ × ½ = ¼ (25%).

The analysis of patterns of inheritance, i.e. the way in which alleles are passed on from one generation to the next, depends on an understanding of these two phenomena:

- Alleles are separated in the production of gametes.
- Alleles combine at fertilisation when gametes fuse.

Different patterns of inheritance, with respect to the alleles of one gene, are provided by:

- dominance — the heterozygote has the same phenotype as the homozygous dominant genotype
- codominance — interaction of alleles results in a heterozygote with its own distinctive phenotype
- lethal allelic combinations — one allele in the homozygous state causes death at an early stage
- multiple alleles — more than two alleles of a gene are possible
- sex linkage — the gene, and its alleles, are located on a sex chromosome, most often the X chromosome

Dominance

Austrian biologist Gregor Mendel carried out breeding experiments in pea plants (*Pisum sativum*) and through these revealed the principles of inheritance. Mendel's work was successful because he bred plants through two generations and in large numbers so that the ratios achieved would be reliable. In one experiment, he crossed white-flowered and purple-flowered plants. He found that the $\mathbf{F_1}$ (first filial generation, or first generation of offspring) were all purple (we now know this is due to a dominant allele) but that when the F_1 were interbred the $\mathbf{F_2}$ (second filial generation) showed a mixture of purple- and white-flowered plants in a ratio of 3:1.

Exam tip

You should develop an awareness of the different ratios possible in genetic crosses. These are often diagnostic of the type of inheritance involved. For example, a 3:1 ratio of phenotypes is caused by two heterozygous individuals with dominance.

In pea plants:

- the allele **P** codes for the production of purple pigment
- the allele **p** does not code for any pigment

The genotypes and phenotypes of flower colour in pea plants are shown in Table 7. The cross is explained in the genetic diagram in Figure 33.

Table 7 The genotypes and phenotypes of flower colour in pea plants

Genotypes	Phenotypes
PP	Purple flowers
Pp	Purple flowers
pp	White flowers

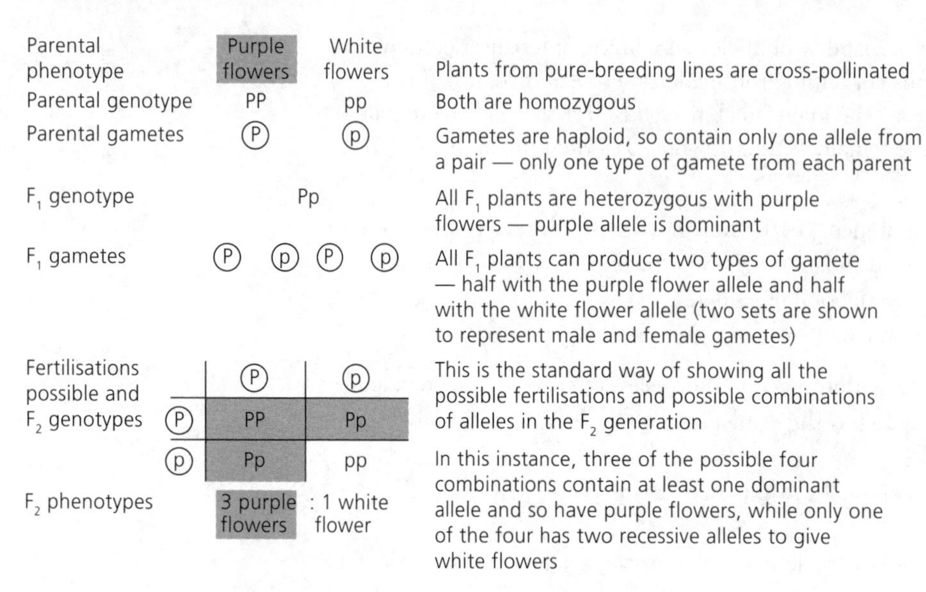

Figure 33 A cross between pure-breeding white-flowered and pure-breeding purple-flowered pea plants

The test cross

Pea plants with purple flowers can be either homozygous (**PP**) or heterozygous (**Pp**). To find out which, a test cross is required. The individual to be tested (e.g. **PP** or **Pp**) is crossed with the homozygous recessive — in this case a plant with white flowers, genotype **pp**. Note that the homozygous recessive can be recognised by its phenotype and can pass on a recessive allele only, which will not influence the phenotype of the offspring. A test cross for purple-flowering plants is shown in Figure 34.

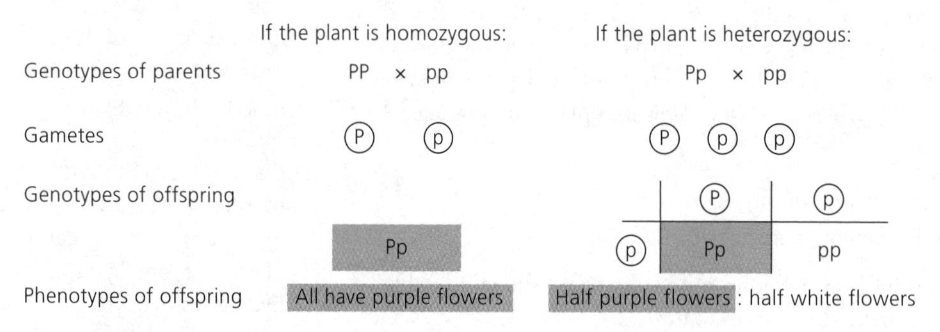

Figure 34 The test cross (testing the genotype of a purple-flowered plant)

If the individual tested is homozygous, all the offspring have the dominant phenotype. If the individual is heterozygous, the offspring will be a mixture of both dominant and recessive phenotypes.

Codominance (incomplete dominance)

In codominance a pair of alleles interacts so that the heterozygote has a phenotype distinctive from (though often intermediate to) each homozygote. Codominance occurs because both alleles are functioning to produce something. An example is provided by pink flower colour in snapdragons (*Antirrhinum*).

Exam tip

Numbers rarely fit a ratio exactly as fertilisation is a random event. To be confident that numbers fit a particular ratio, it is necessary to use the chi-squared (χ^2) test to determine whether numbers are significantly different from those expected.

Knowledge check 61

When a straight-winged fruit fly (*Drosophila*) was crossed with another straight-winged fly, the offspring were a mixture of straight-winged and curled-winged flies. Explain why the straight-winged offspring would be a mixture of homozygous and heterozygous flies. Which type of fruit fly would be used in test crosses?

In snapdragons:

- the allele $\mathbf{C^R}$ codes for the production of red pigment
- the allele $\mathbf{C^W}$ codes for the production of white pigment

The genotypes and phenotypes of flower colour in snapdragons are shown in Table 8.

Table 8 The genotypes and phenotypes of flower colour in snapdragons

Genotypes	Phenotypes
C^RC^R	Red flowers
C^RC^W	Pink flowers
C^WC^W	White flowers

A cross between two heterozygotes produces a distinctive 1:2:1 ratio. This is shown in a cross between two pink-flowered snapdragons, explained in the genetic diagram in Figure 35.

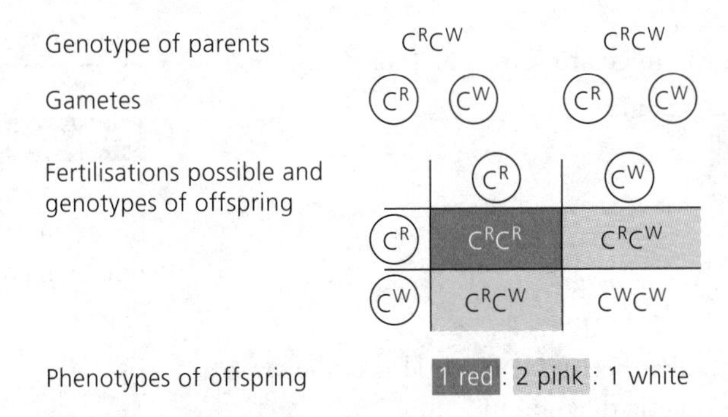

Figure 35 A cross between two pink-flowered snapdragons

Knowledge check 62

Two heterozygotes for a particular condition produce offspring in the expected ratio 1:2:1. What type of inheritance does this show?

Lethal allelic combinations

Alleles that determine flower colour may influence the attractiveness of the flower to a pollinating insect but they are not in themselves essential for the metabolism of the plant. However, many alleles code for essential proteins (e.g. enzymes, membrane carriers or membrane receptors) and the absence of such proteins prevents the operation of vital functions. A deviant of such an allele is called a lethal allele. In general, the 'normal' allele has only to be present once (i.e. even in the heterozygous condition) for the protein to be produced and the vital function to be carried out. (Note that it is *normal* for alleles to differ and so the 'normal' allele is often called the wild-type allele in genetics.) Thus, most lethal alleles are recessive and so the lethal phenotype occurs only in individuals homozygous for the allele.

Different types of lethality are recognised:

- The lethal allelic combination causes death of either the zygote or an early embryonic stage so there is no obvious evidence that the lethal allelic combination ever occurred.
- The lethal allelic combination causes death after a reduced lifespan. For example, Tay–Sachs disease is a rare genetic disorder of the central nervous system in humans. It is caused by the absence of the enzyme hexosaminidase (Hex-A). Without Hex-A, a lipid called GM2 ganglioside accumulates abnormally in cells,

particularly in the nerves of the brain. The ongoing accumulation results in progressive damage to the nerve cells, and death occurs in early childhood.

- The lethal allelic combination causes death at an early stage of development but its presence is evident in the heterozygote where it displays a distinctive phenotype. One example is provided by yellow coat colour in mice: the wild-type allele (**A**) codes for the production of a signalling protein involved in a range of functions, most noticeably the production of yellow and black bands on individual hairs (giving the agouti coat colour); it is also associated with proper embryonic development. The $\mathbf{A^Y}$ allele fails to code for this protein and is associated with obesity, diabetes and increased susceptibility to cancer, and a yellow coat colour from a failure to produce black pigment. The homozygous state, $\mathbf{A^YA^Y}$, is lethal since it lacks the **A** allele for proper embryonic development.

In mice:

- the allele **A** codes for a signalling protein
- the allele $\mathbf{A^Y}$ does not code for this protein

The genotypes and phenotypes of coat colour in mice are shown in Table 9.

Table 9 The genotypes and phenotypes of coat colour in mice

Genotypes	Phenotypes
AA	Agouti coat colour
AA^Y	Yellow coat colour
A^YA^Y	Lethal: embryonic death

A cross between two heterozygotes produces a distinctive 2:1 ratio. This is shown in a cross between yellow mice, explained in the genetic diagram in Figure 36.

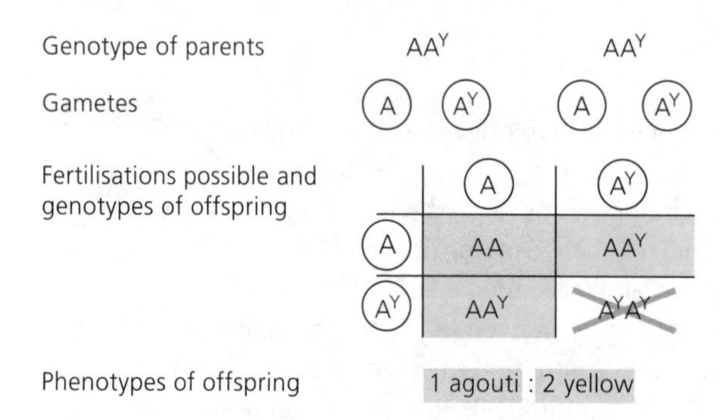

Figure 36 A cross between yellow mice

Multiple alleles

Multiple alleles describes the situation where more than two alleles of a particular gene exist in the population (of course only two can exist in any one individual). An example is the inheritance of the ABO blood group, which also demonstrates codominance and dominance: two alleles are active in coding for the production of the A-antigen and the B-antigen and so are codominant; a third allele does not code for any antigen and so is recessive to the other two alleles. The genotypes and phenotypes are shown in Table 10.

Knowledge check 63

Dominant lethal alleles are rarely detected. One exception is the dominant allele in Huntington's disease, a neurological disorder in humans which reduces life expectancy. In general, symptoms develop after the age of 40. Suggest (a) why dominant lethal alleles are rarely detected, and (b) how Huntington's disease is an exception.

Knowledge check 64

In the Mexican hairless breed of dogs, the hairless condition is produced by the heterozygous genotype (**Hh**). Normal dogs are homozygous **HH**. Puppies homozygous for the **h** allele are usually born dead. What would be the phenotypic ratio of hairless and normal offspring from matings between hairless and normal dogs?

Table 10 The genotypes and phenotypes of blood groups in humans

Genotypes	Phenotypes
I^AI^A, I^AI^o	Blood group A
I^BI^B, I^BI^o	Blood group B
I^AI^B	Blood group AB
I^oI^o	Blood group O

In blood group inheritance in humans:

- the allele $\mathbf{I^A}$ codes for the production of antigen A
- the allele $\mathbf{I^B}$ codes for the production of antigen B
- the allele $\mathbf{I^o}$ does not code for any antigen

A cross between a parent heterozygous for blood group A and a parent heterozygous for blood group B has the potential to produce children with any of the four blood groups. This is shown in Figure 37.

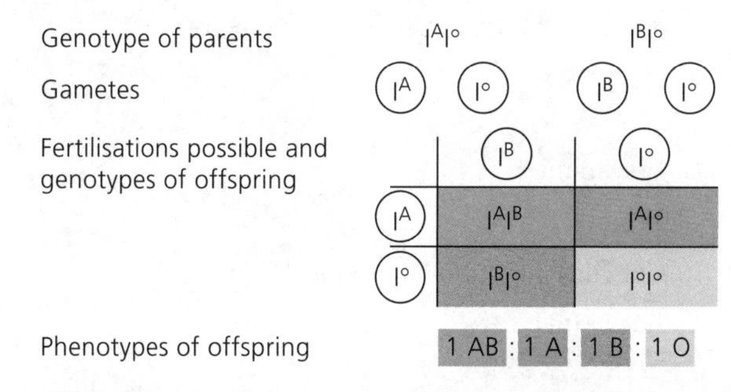

Figure 37 A cross between a parent heterozygous for blood group A and a parent heterozygous for blood group B

Another well-known example of multiple alleles is provided by fur type in rabbits. There are four alleles with a dominance hierarchy:

- **C** — dominant to $\mathbf{c^h}$, $\mathbf{c^{ch}}$ and $\mathbf{c^a}$
- $\mathbf{c^h}$ — dominant to $\mathbf{c^{ch}}$ and $\mathbf{c^a}$
- $\mathbf{c^{ch}}$ — dominant to $\mathbf{c^a}$
- $\mathbf{c^a}$ — recessive

The genotypes and phenotypes for fur type in rabbits are shown in Table 11.

Table 11 The genotypes and phenotypes of fur type in rabbits

Genotypes	Phenotypes
CC, Cc^h, Cc^{ch}, Cc^a	Agouti fur
c^hc^h, c^hc^{ch}, c^hc^a	Himalayan fur
$c^{ch}c^{ch}$, $c^{ch}c^a$	Chinchilla fur
c^ac^a	Albino fur

Only a rabbit with albino fur has a known genotype — $\mathbf{c^ac^a}$. Others would require a test cross to ascertain their genotypes, i.e. cross the individual with the homozygous recessive, $\mathbf{c^ac^a}$ (an albino rabbit).

Knowledge check 65

Four babies were born with the blood groups A, B, AB and O. The parents were known as W, X, Y and Z and had the following blood groups: parents W had blood groups A and B; X had B and O; Y had O and O; Z had AB and O. Which baby belonged to which parents?

Exam tip

In an exam you may be asked about the inheritance of a trait determined by multiple alleles. Remember that any individual can possess only two alleles and that only one of these may be passed on via a gamete to be combined with a single allele in a gamete from the other parent.

Sex linkage

In mammals, one pair of chromosomes is associated with gender. These are the sex chromosomes, designated **XX** in females and **XY** in males. (The other, non-sex chromosomes are called autosomes.) All eggs contain an X chromosome, half of the sperm contain an X chromosome and half contain a Y chromosome.

The X and Y chromosomes are largely non-homologous and so carry different genes. Indeed, the Y chromosome carries very few genes and these are, in general, concerned with the development of maleness — for example, the gene *SRY* triggers testis development. Sex-linked inheritance occurs with the alleles of genes located on the sex chromosomes. **Y-linked inheritance** is confined to males and is relatively rare. **X-linked traits** are relatively common.

Fruit flies (*Drosophila melanogaster*) also have XX females and XY males. An example of an X-linked recessive trait is 'white eye'. Genotypes of sex-linked traits include the appropriate chromosome, X or Y, as well as the allele. The dominant wild-type allele, $\mathbf{X^+}$, determines red eye, while $\mathbf{X^W}$ represents the recessive white eye allele. There is no equivalent allele on the Y chromosome.

In fruit flies:

- the wild-type allele $\mathbf{X^+}$ codes for red eye and is dominant
- the allele $\mathbf{X^W}$ codes for white eye and is recessive

The genotypes and phenotypes for eye colour in fruit flies are shown in Table 12.

Table 12 The genotypes and phenotypes of eye colour in fruit flies

Genotypes	Phenotypes
X^+X^+, X^+Y	Red-eyed female and red-eyed male
X^+X^W	Red-eyed female (carrier)
X^WX^W, X^WY	White-eyed female and white-eyed male

Figure 38 shows reciprocal crosses between red-eyed and white-eyed fruit flies:

- a red-eyed female and a white-eyed male
- a white-eyed female and a red-eyed male

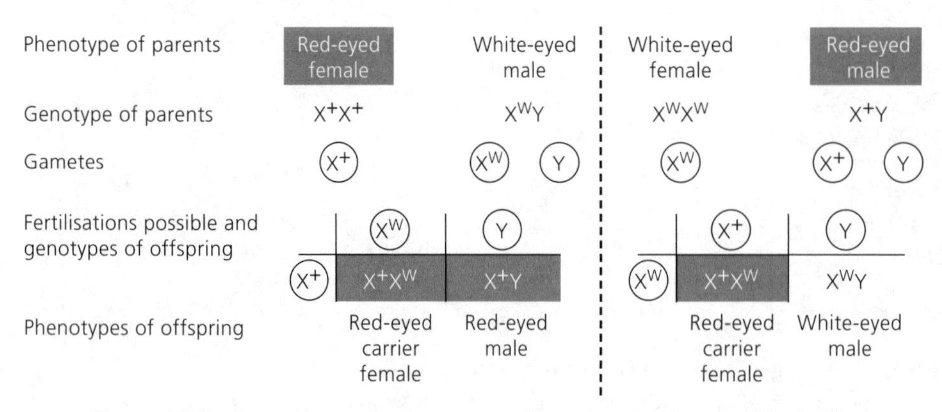

Figure 38 Reciprocal crosses between red-eyed and white-eyed fruit flies

Exam tip

You should be aware that human karyotypes are used for chromosome analysis. Pair 23 constitutes the sex chromosomes that indicate gender; an extra (third) chromosome number 21 indicates Down syndrome.

Exam tip

In sex linkage, the sex chromosomes XX and XY are always shown and the dominant or recessive alleles of particular genes are represented by superscript upper- and lower-case letters. Thus in the inheritance of haemophilia, $\mathbf{X^HX^h}$ represents a female carrier and $\mathbf{X^hY}$ an affected male — note that the Y chromosome is 'barren'.

X-linked recessive traits share the following features:

- Reciprocal crosses produce different results and are used in testing for sex linkage.
- They are more common in males than females. This is because the allele has to appear only once in the male but has to be inherited from both parents in an affected female.
- Affected males inherit the allele from the female parent since the Y chromosome is inherited from the male parent (and most commonly from a heterozygous — carrier — female parent).
- Affected females inherit the allele from both parents (and so the male parent must also be affected).

There are a number of X-linked recessive traits in humans — for example, red–green colour blindness and haemophilia. In cats there is the interesting example of a coat colour called tortoiseshell. The condition is codominant and so tortoiseshell cats are heterozygotes. It occurs in females only since males cannot have both alleles.

X-linked dominant traits exist, such as hereditary hypophosphatemic rickets in humans. The trait is passed from an affected father to all his daughters (since a daughter always inherits her father's X chromosome).

The sex-determination system in mammals is not universal. For example, in birds, butterflies and moths the female is XY and the male is XX. In alligators and crocodiles it is the temperature at which eggs develop that determines gender. Most eggs hatched between 26°C and 30°C become female; temperatures between 34°C and 36°C produce mostly males.

Human genetics and pedigrees

A pedigree charts the transmission of a genetic trait over several generations in a family. Pedigrees can be used to analyse the pattern of inheritance of genetic disorders. Two pedigrees are illustrated in Figure 39.

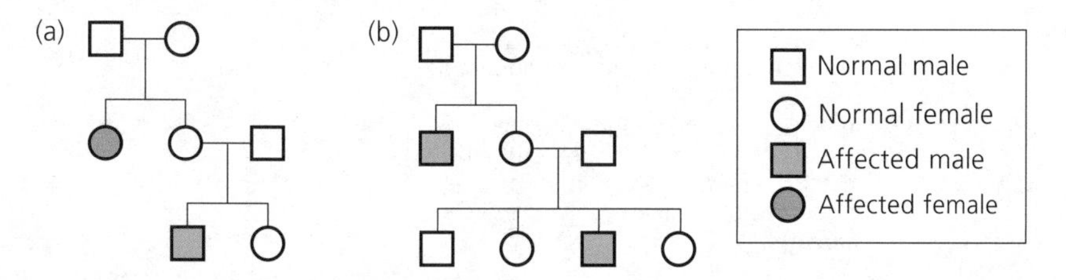

Figure 39 Pedigrees for (a) cystic fibrosis, and (b) Duchenne muscular dystrophy

Both traits are recessive — affected individuals have parents without the trait. Cystic fibrosis is autosomal recessive — it cannot be X-linked since an affected female has a father who is unaffected (see Figure 39a). Duchenne muscular dystrophy is X-linked recessive — Figure 39(b) shows that the trait is more common in boys. However, a more extensive pedigree would be required to determine X-linkage with any degree of certainty.

Considerations when analysing pedigrees include:

- **autosomal recessive** — parents of an affected individual may not be affected, i.e. the trait may appear to skip a generation

Knowledge check 66

Would it be possible for a girl to have haemophilia? Explain your answer. What would be the notation for a female haemophiliac?

Exam tip

If you are given a question on the inheritance of, say, feather colour in budgerigars, you will be told that males are homogametic (XX) and females are heterogametic (XY). This is another example of the need to read questions carefully and use the information provided.

Exam tip

If you are asked to solve the genetics of a pedigree (as in Figure 39) you should look for clues about the type of inheritance. For example, if two parents with the same phenotype have a child of differing phenotype, then the parents are heterozygous and the child is homozygous recessive. Note these genotypes on the figure and work from there.

- **X-linked recessive** — as for autosomal recessive but:
  - more common in males than females (passed on from carrier mothers)
  - for a female to be affected the father must also be affected
  - an affected female will pass the trait on to all her sons
- **autosomal dominant** — at least one parent must be affected (though if both parents are affected not all the children would necessarily be affected, since the parents could be heterozygous)
- **X-linked dominant** — as for autosomal dominant but an affected father will pass on the trait to all his daughters

Exam tip

In an exam question you could be expected to state either of Mendel's first and second laws and explain it with respect to the events in meiosis. The separation of homologous chromosomes during anaphase I explains the 'segregation of alleles'; the random array of homologous chromosome pairs on the spindle during metaphase I explains 'independent assortment'.

Heredity of alleles at two loci: dihybrid inheritance

Mendel also performed crosses in which he followed the segregation of the alleles of two genes. These experiments formed the basis of **Mendel's second law**, the law of independent assortment, which states that '*during gamete formation, the segregation of the alleles of one gene is independent of the segregation of the alleles of another gene*'. This is explained by the random arrangement of the homologous pairs on the equator of the spindle at metaphase I of meiosis and their subsequent separation during anaphase I.

Dominance at each of the two loci

Figure 40 shows the results of a cross between a pure-breeding tall, purple-flowered pea plant with a pure-breeding short, white-flowered pea plant through to the F_2 generation.

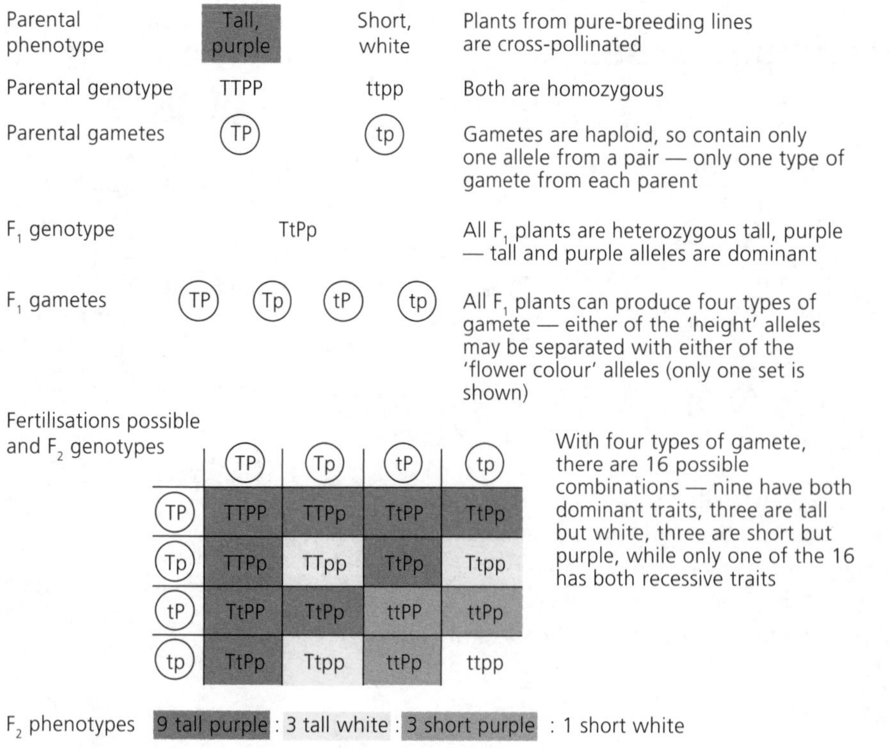

Parental phenotype	Tall, purple	Short, white	Plants from pure-breeding lines are cross-pollinated
Parental genotype	TTPP	ttpp	Both are homozygous
Parental gametes	TP	tp	Gametes are haploid, so contain only one allele from a pair — only one type of gamete from each parent
F_1 genotype	TtPp		All F_1 plants are heterozygous tall, purple — tall and purple alleles are dominant
F_1 gametes	TP Tp tP tp		All F_1 plants can produce four types of gamete — either of the 'height' alleles may be separated with either of the 'flower colour' alleles (only one set is shown)

Fertilisations possible and F_2 genotypes

	TP	Tp	tP	tp
TP	TTPP	TTPp	TtPP	TtPp
Tp	TTPp	TTpp	TtPp	Ttpp
tP	TtPP	TtPp	ttPP	ttPp
tp	TtPp	Ttpp	ttPp	ttpp

With four types of gamete, there are 16 possible combinations — nine have both dominant traits, three are tall but white, three are short but purple, while only one of the 16 has both recessive traits

F_2 phenotypes 9 tall purple : 3 tall white : 3 short purple : 1 short white

Figure 40 A cross between pure-breeding tall, purple-flowered and pure-breeding short, white-flowered pea plants

Independent assortment works only for genes located on different chromosomes. However, chromosomes, particularly those that are longer, are composed of many genes — a genetic situation called linkage.

Exam tip

You are not required to solve problems involving linkage, i.e. the inheritance of genes on the same chromosome. However, you should be aware that humans have approximately 21 000 genes present on just 23 pairs of chromosomes — so any single gene may be linked with a thousand other genes.

Knowledge check 67

In budgerigars, body feathers may be green or blue (**G** = dominant 'green body' allele, **g** = 'blue body' allele), while the stripes on the back and wings may be black or brown (sex-linked alleles, $\mathbf{X^B}$ = dominant 'black stripes' allele, $\mathbf{X^b}$ = 'brown stripes' allele). Females are XY and males XX. What would be the genotype, phenotype and gender of offspring resulting from the cross: $\mathbf{GgX^BY}$ × $\mathbf{ggX^bX^b}$?

Different genetic situations

In the situation described above, there were dominant alleles at each genetic locus. However, codominance, lethal allelic combinations, multiple alleles and sex linkage are all possible — for example, there may be codominance at one genetic locus and sex linkage at another.

Polygenic inheritance: the additive effect of alleles at different loci

Many traits are governed by the cumulative effects of two or more genes. For example, grain colour in wheat (*Tritium aestivum*) is determined by the additive effects of the alleles of two genes, **A/a** and **B/b**. The more **A** and **B** alleles that are present in a genotype, the deeper the shade of red in the grain. The absence of **A** and **B** alleles causes the grain to be white. Indeed, five shades of colour are possible: darkest red, dark red, red, light red and white. The results of a cross between two individuals of genotype **AaBb**, red grain colour, are shown in Figure 41.

Parental phenotypes	Red grain	Red grain	Plants with the intermediate colour of grain
Parental genotypes	AaBb	AaBb	Both are heterozygous
Parental gametes	(AB) (Ab) (aB) (ab)	(AB) (Ab) (aB) (ab)	Each produces four types of gamete — A and a are segregated with either of B and b

Fertilisations possible and F2 genotypes

Phenotypes of offspring

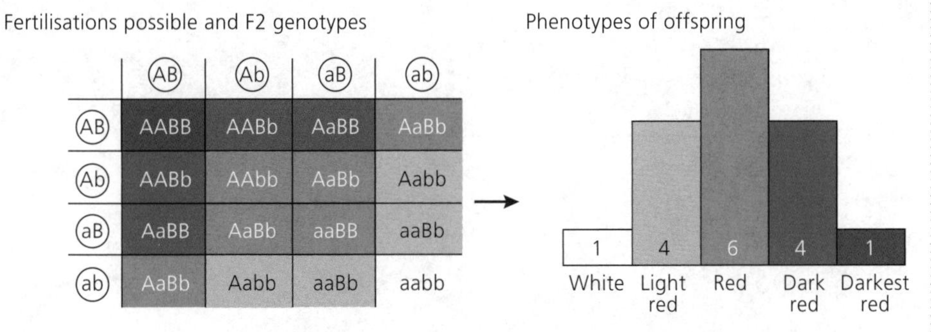

Figure 41 A cross between two red-grained wheat plants of genotype AaBb

Where more than two genes are involved, **polygenic inheritance** results. This explains quantitative characteristics (such as height or weight) that exhibit the normal frequency distribution for a continuous variable, particularly when the further influence of the environment is added.

Epistasis: the interference of gene expression by the alleles at another locus

Epistasis is the situation where one gene interferes with the expression of another. For example, in squash plants fruit colour is determined by the interaction of two genes, **W/w** and **G/g**. The **W** allele suppresses the action of the **G/g** gene so the fruit are white. When the genotype is **ww**, the **G/g** gene is expressed, with the dominant **G** allele producing yellow fruit colour. The **g** allele is expressed recessively as green. A cross between two squash plants of genotype **WwGg** produces an offspring genotypic ratio of 9 **W–G–**:3 **W–gg**:3 **wwG–**:1 **wwgg**, and so the phenotypic ratio is 12 white:3 yellow:1 green.

There are many examples of epistasis. Indeed, a purple flower in pea plants involves the action of two enzymes each produced by a different gene:

$$\underset{\text{(colourless)}}{\text{Precursor}} \xrightarrow{\text{gene A}} \underset{\text{(colourless)}}{\text{Intermediate}} \xrightarrow{\text{gene B}} \text{Purple pigment}$$

A cross between two purple-flowered plants of genotype **AaBb** would produce an offspring genotypic ratio of 9 **A–B–**:3 **A–bb**:3 **aaB–**:1 **aabb** and so the phenotypic ratio is 9 purple:7 white.

A further example is provided by coat colour in mice. The agouti (banded hair) allele, **A**, is dominant to the black allele, **a**. The coat colour alleles are expressed only if the **E** allele of an epistatic gene is present, i.e. they are not expressed if the genotype is **ee** and a white coat is produced. A cross between two agouti mice of genotype **AaEe** would produce an offspring genotypic ratio of 9 **A–E–**:3 **aaE–**:3 **A–ee**:1 **aaee** and so the phenotypic ratio is 9 agouti:3 black:4 white.

Knowledge check 69

An agouti mouse (genotype **AaEe**) is crossed with a homozygous white mouse (genotype **aaee**). The **A/a** gene is expressed only if the **E** allele of an epistatic gene is present (**e** represents its recessive allele). What genotypes and phenotypes would you expect in their offspring, and in what ratio?

Knowledge check 68

How many types of gametes would be produced by a trihybrid, **AaBbCc**? List them.

Exam tip

Note that polygenic inheritance and epistasis are both examples of gene interaction — they involve two (or more) genes that influence the same characteristic.

Skills development

Solving genetics problems

Genetics problems are generally presented in one of two ways:

- You are presented with details about the alleles involved, given a cross and asked to work out the genotypes and phenotypes of the offspring.
- You are presented with the results of a cross and asked to work out details of the alleles (and genes) involved.

Table 13 shows some tips for solving genetics problems presented in either way.

Table 13 Tips for solving genetics problems

When presented with details about the alleles involved	When presented with the offspring phenotypes resulting from a cross
You may be presented with the alleles (and the symbols to use), whether two genes are involved, and details such as whether sex linkage is involved. The questions that follow will involve working out the genotypes and phenotypes resulting from a particular cross. ■ Write out the alleles and the genotypes involved (using distinctive upper- and lower-case letters, e.g. **A** and **a**, though these are usually given in the question). ■ Write out the phenotypes involved. The initial question may well ask you to work out the possible genotypes of particular phenotypes. ■ Present the cross in a genetic diagram, as shown in the previous figures: – **Parents** — show parental genotypes (and phenotypes). – **Gametes** — show the types of gametes produced, i.e. apply the laws of inheritance. Only one allele of the pair enters a gamete. If homozygous then there is only one type of gamete; if heterozygous two types are produced. Either allele of one gene may be separated with either allele of another gene. If homozygous at both genes then there is one type of gamete; if homozygous at one gene and heterozygous at the other, two types are produced; if heterozygous at both loci then four types are produced. – **Fertilisations** — use a Punnett square if there are two (or more) *types* of gamete from each parent: a 4 × 4 Punnett square if both parents produce four types of gametes, or a 2 × 4 Punnett square if one parent produces two types of gamete while the other produces four types. You do *not* need to use a Punnett square if only one type of gamete is produced, even by just one parent. – **Offspring** — summarise the offspring genotypes and phenotypes and their proportions. It can help if you add the phenotypes under the offspring genotypes in your Punnett square.	The questions that follow will involve working out the genetics of the situation. For example, which allele is dominant, is there evidence of lethality or sex linkage? ■ It can help to summarise the information given, e.g.: taster × taster → tasters and non-tasters. ■ Study the information: pure-breeding individuals are homozygous; if they do not breed true they are heterozygous. You can then work out which allele is recessive. ■ A phenotype that exhibits the recessive trait is homozygous. ■ If the numbers of offspring allow, use the *phenotypic ratios* to help you. They can be distinctive: – **With the alleles of one gene** — a 3:1 ratio indicates two types of gamete from each of two heterozygous parents; 1:1 indicates two types of gamete from one heterozygous parent and one type from the other, homozygous parent; 1:2:1 indicates codominance with both parents heterozygous; 2:1 is indicative of a lethal combination where both parents are heterozygous. – **With the alleles of two genes** — a 9:3:3:1 results from a 4 × 4 Punnett square, so both parents must be heterozygous at each locus and produce four types of gamete; this is also the case where any ratio 'adds' up to 16 (such as 9:7, 9:3:4 or other 'odd' ratios indicating gene interaction); 3:1:3:1 results from a 2 × 4 Punnett square, so one parent is heterozygous at one locus and homozygous recessive at the other and the other parent is heterozygous at both; 1:1:1:1 may result from a 2 × 2 Punnett square where one parent is heterozygous at one locus and homozygous recessive at the other, while the other parent has the reverse situation (e.g. **Aabb** × **aaBb**), or may result from a 1 × 4 Punnett square where one parent is homozygous recessive at both loci and the other parent is heterozygous at both loci (e.g. **aabb** × **AaBb**).

Summary

- Diploid organisms possess pairs of homologous chromosomes that carry alleles of the same gene at the same locus. If the alleles are the same, the individual is homozygous; if the alleles are different, he/she is heterozygous.
- There are different patterns of inheritance:
  - dominance — dominant alleles are always expressed; recessive alleles are expressed only when homozygous
  - codominance — two alleles are both expressed so that the heterozygote has a distinctive phenotype
  - lethality — one homozygote may inherit a lethal combination, so that two heterozygotes produce a 2:1 phenotypic ratio
  - multiple alleles — there are more than two alleles of a gene, so there is more than one type of heterozygote
  - sex linkage — alleles are located on a chromosome that determines gender (e.g. an X chromosome in mammals)
- A test cross determines whether an organism showing a dominant feature is homozygous or heterozygous — the organism is bred with one that shows the recessive trait.
- Dihybrid inheritance involves two separate genetic loci. If both parents are heterozygous for the two genes and, at each locus, an allele exhibits dominance, then a 9:3:3:1 phenotypic ratio results.
- Polygenic inheritance occurs when the alleles of many different genes contribute to a characteristic.
- Epistasis occurs when an allelic combination at one locus determines whether the alleles at another locus are expressed.
- Mendel's laws of inheritance are:
  - the law of segregation (first law) — only one allele is inherited via a gamete
  - the law of independent assortment (second law) — an allele of one gene may be inherited with either allele of another gene

Population genetics, evolution and speciation

Population genetics

A population is a group of individuals of the same species living in the same habitat because they are similarly adapted. Members of the population are capable of interbreeding and so their genes and alleles are free to mix. All the genes and alleles in a population at a particular time make up the **gene pool**. The proportion of a particular allele in the population is called the **allele frequency**.

Exam tip

Population genetics considers the alleles of a single gene in an entire population — the gene pool — with one of two alleles dominant.

Consider a single gene that has two alleles — **A** and **a** — in the population. The frequencies of the alleles, **A** and **a**, are represented by the algebraic symbols p and q, respectively. Since **A** and **a** are the only alleles of that gene, their combined frequencies must equal 1:

$$p + q = 1$$

Genotypes are formed from the fusion of gametes, each of which contains a single allele, **A** or **a**. If the probability of the different alleles combining at fertilisation depends solely on their respective frequencies in the gene pool, then the **genotype frequencies** are determined as:

genotype AA has a frequency of p^2

genotype aa has a frequency of q^2

genotype Aa has a frequency of $2pq$

The condition for allowing this determination is often described as **random mating** (with respect to the alleles in question). Again, since these are all the genotypes in the population, their sum must equal 1:

$$p^2 + 2pq + q^2 = 1$$

This is called the Hardy–Weinberg equation.

The Hardy–Weinberg equation relates allele frequency to genotype frequency, i.e. it allows for the determination of one from the other. If the allele frequencies (p and q) are known, then the genotype frequencies can be determined. It is more likely, however, that the phenotype frequencies are known (since these are what can be seen). Even then, the equations can be used to determine the frequencies of the alleles and of the different genotypes. For example, the cystic fibrosis (*CF*) gene has two alleles, the normal **F** allele that leads to normal mucus production and the recessive **f** allele that leads to the production of thicker mucus and, hence, cystic fibrosis. The pair of alleles of the *CF* gene has three possible combinations — homozygous dominant (**FF**), homozygous recessive (**ff**) and heterozygous (**Ff**). Both **FF** and **Ff** (carriers) have the normal phenotype; **ff** has cystic fibrosis. In Ireland, the incidence of cystic fibrosis is around 1980 in a total population of 6.6 million. This represents a frequency of 0.0003 (or 0.03%, though it is better to express frequencies as decimal fractions because the arithmetic is more straightforward). If the frequency of the homozygous recessive, **ff**, is equal to q^2 then:

$q = \sqrt{0.0003} = 0.017$

$p = 0.983$ (since $p + q = 1$)

frequency of **FF** = $p^2 = 0.966$

frequency of **Ff** = $2pq = 0.033$

So the frequency of those who carry the **f** allele in Ireland is 0.033. This might seem small but it means that 0.033 multiplied by 6.6 million, i.e. 217 800, is the number who carry the **f** allele. Viewed locally, it means that in a school population of 1000, 33 students might be expected to be carriers.

Knowledge check 70

The banded snail (*Cepaea nemoralis*) is common in sand dunes where the colour of the shell is either yellow or pink. The allele for pink shell is dominant. In a particular population of snails, 576 had pink shells and 1024 had yellow shells. Using the Hardy–Weinberg equation, calculate the number of snails expected to be heterozygous.

Exam tip

In the Skills section you are directed to work in decimals and *not* in percentages. The reason for this is simple and is illustrated by the following example: if q^2 = 36%, then q = 60% (not 6%!) since 36% is 36 ÷ 100 and its square root equals 6 ÷ 10 = 0.6 (or 60%). Students often have difficulty working with percentages, so use decimals (divide the percentage by 100).

Skills development

Calculating allele and genotype frequencies using the Hardy–Weinberg equation

The use of the Hardy–Weinberg equation to calculate genotype frequencies (p^2, $2pq$ and q^2) is simple if you are given the allele frequencies (p and q). However, it is more likely that any question will present you with the numbers of phenotypes in a population. You must remember that the dominant trait consists of both homozygotes and heterozygotes. Only the homozygous recessive individuals contribute towards the recessive trait.

frequency of dominant trait = $p^2 + 2pq$
frequency of recessive trait = q^2

It is difficult to work arithmetically with percentages, so frequencies must always be presented as proportions, i.e. as decimal fractions. The steps in solving problems are:

1. Determine the frequency of the recessive trait. Ensure that it is presented as a proportion (decimal fraction). This is q^2.
2. Calculate q (as the square root of q^2). This is the frequency of the recessive allele (e.g. *a*).
3. Calculate p (as $p = 1 - q$). This is the frequency of the dominant allele (e.g. *A*).
4. Calculate p^2. This is the frequency of the homozygous dominant individuals in the population. The actual number of homozygous dominant individuals in the population can be calculated as $p^2 \times N$ (where N is the total number of individuals in the population).
5. Calculate $2pq$. This is the frequency of the heterozygotes in the population. The actual number of heterozygous individuals in the population can be calculated as $2pq \times N$.

Exam tip

To calculate q^2 you need to know the number with the recessive trait and the total number in the population. However, examiners may give you the number with the dominant trait and the total — subtract these to get the number of recessives. Or they may give you the numbers with the dominant and recessive traits — add these to get the total number.

While the Hardy–Weinberg equation is useful for studying genes in populations, it must be remembered that its use entails certain assumptions. The **Hardy–Weinberg principle** is that no factor operates that would cause the allele frequencies in the offspring generation to differ from those in the parental population. This means that allele frequencies remain constant over time — the population is said to be in Hardy–Weinberg equilibrium. The conditions for this to happen include the following:

- *Mating is random.* In fact, non-random mating causes changes in the proportions of genotypes without affecting allele frequencies — for example, inbreeding promotes homozygosity (more alleles than expected appear in the homozygous state).
- *The population is large.* In small populations random changes can disrupt the allele frequencies, a situation called genetic drift.
- *No mutations are occurring.* In general, the rate of gene mutation is so low (in the order of 1 in 10 million) that it has little impact on allele frequencies. (The significance of mutation is the introduction of new alleles into populations.)
- *No migration is taking place* (either into or out of the population).
- *No selection is taking place* (i.e. all alleles are equally advantageous). Selection is the principal agent influencing changes in the frequency of alleles in populations.

If it is found that the allele frequencies of a gene pool are changing over time, then it is evidence that selection is operating in the population.

Exam tip

You must distinguish between the Hardy–Weinberg equation and the Hardy–Weinberg principle. The Hardy–Weinberg equation shows genotype frequencies described by p^2, $2pq$ and q^2, and that their sum is equal to 1 ($p^2 + 2pq + q^2 = 1$). The Hardy–Weinberg principle is that, provided certain assumptions hold, the genotype frequencies remain constant over time.

Selection

Selection and genetic variation

Populations are genetically variable. There are a number of reasons for this:

- **Mutation** is a source of genetic variation. **Gene mutation** introduces new alleles into the population. **Chromosome mutation** occurs when chromosomes fail to separate during meiosis so that extra (or occasionally fewer) chromosomes appear in the affected individual.

- **Meiosis** (whether in the production of gametes in animals or in the production of spores in plants) produces a variety of haploid cells through independent assortment and recombination.
- **Sexual reproduction**, particularly when cross-fertilisation is ensured, is the most important factor promoting genetic variability in populations. It involves the mixing of genetic material from the haploid cells of two different individuals. In life cycles, meiosis (production of haploid cells) and sexual reproduction act to shuffle the allelic combinations.
- Variability is preserved by diploidy, which shelters rare, recessive alleles, i.e. heterozygotes act as important reservoirs of genetic variation in populations.

In a genetically variable population, some individuals will possess certain alleles, or combinations of alleles, that make them better adapted to a particular environment. These individuals survive and reproduce more successfully than others in that environment. This is **natural selection**, which is defined as *'the unequal transmission of alleles to subsequent generations by different genotypes'*.

Exam tip

Selection acts on the natural genetic variation in a population. For example, if DDT is applied to a population of mosquitoes then one in a million may be resistant to it and so survives, reproduces and passes on resistance. Do *not* say that DDT causes mutation. Mutation is a random event and will have happened long before the application of DDT.

Polymorphic populations may be used to study natural selection. Polymorphism is the presence in a population of two or more distinct forms. For example, the snail *Cepaea nemoralis* possesses shells that vary in colour (brown, pink or yellow) and in the number of bands (five, three or none). Studies have shown that thrush predation is a powerful agent of selection: in grassland, yellow, banded forms are selected for as they are better camouflaged; on the floors of beech forest, brown, unbanded snails are favoured. The result is considered to be a balanced polymorphism.

Different forms can also be maintained in a population by a type of selection called apostatic selection — a predator hunts the most common form (since it is easier to see) until it becomes the less common form. At this point, another form becomes the favoured prey. Heterozygous advantage is sometimes suggested as a means of maintaining different forms in a population. For example, if **Aa** was selected for (and both **AA** and **aa** disadvantaged) then, while **Aa** would increase in frequency, it would always interbreed to produce more homozygotes.

Natural selection can act to remove some of the variants from a population and so reduce the amount of genetic variation. There are different ways in which selection might do this, including stabilising selection and directional selection.

Exam tip

You should be aware that mutation is generally deleterious — for example, a chromosome mutation leading to an extra chromosome 21 results in Down syndrome.

Exam tip

Remember the **fitness** of an organism is its ability to pass on alleles to subsequent generations. The fittest (best adapted) individual in a population is the one that produces the largest number of offspring (which themselves survive to reproduce).

Knowledge check 71

(a) What mechanisms generate genetic variation in a sexually reproducing population? (b) What is the effect of selection on the range of genetic variation in a population?

Stabilising selection

Stabilising selection occurs *where environmental conditions are largely unchanging.* It favours the modal or intermediate forms and acts against the extremes.

An example of stabilising selection comes from human birth records of babies born in London between 1935 and 1946 (see Figure 42). It shows that there was an optimum birth weight for babies and that babies with birth weights heavier or lighter were at a selective disadvantage.

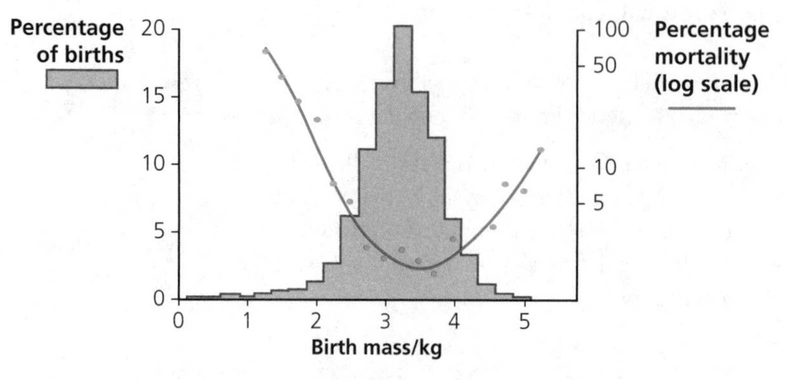

Figure 42 Stabilising selection in size of human babies

Stabilising selection does not lead to evolution — it maintains the **adaptive norm**.

Exam tip

Studies on the wing length of house sparrows (*Passer domesticus*) have shown that those with wing lengths at or near the mean tend to survive adverse weather conditions better than those with longer or shorter wings.

Directional selection and evolutionary change

Directional selection is associated with *changing environmental conditions*. In these situations, the majority of an existing form may no longer be best suited to the environment. Some extreme forms of the population may have a selective advantage (i.e. be more likely to survive and breed) in the changed conditions. They will therefore contribute more offspring, and the alleles these offspring possess, to the next generation. The result is a change in the genetic composition of the population. This is **evolution**, the basis of which is a change in allele frequency in the population.

The classic example of directional selection is provided by the changes in the proportions of light and dark (melanic) forms of the peppered moth (*Biston betularia*), as a result of industrial pollution. Before industrial pollution the light form (genotype **cc**) was camouflaged on the lichen-covered trunks of trees; the dark form (with the dominant allele **C**) was selected against and occurred only by recurrent mutation. Pollution killed the lichens and tree trunks became blackened with soot. Therefore, the rare dark form was selected for and, in Manchester, its frequency in the population increased to 98%.

Knowledge check 72

What effect did industrial pollution have on: (a) the frequency of the **C** (melanic) allele in a population of peppered moths, (b) the rate of mutation of the **c** allele to the **C** allele?

Knowledge check 73

Artificial selection is the intentional breeding of animals and plants for certain desirable traits — for example, increased milk production in cows. What type of selection does artificial selection resemble?

Speciation

Speciation is the evolution of new species from ancestral species. A **species** is defined as *a group of organisms with similar morphological, physiological, biochemical and behavioural features that can interbreed to produce fertile offspring, and which is reproductively isolated from other species.* New species evolve when genetic differences develop that prevent them from freely interbreeding.

Speciation through isolation: allopatric speciation

Allopatric speciation is the evolution of new species as a result of **geographical isolation**. The process takes place as follows and is shown in Figure 43:

- The ancestral species expands its range into new locations, although the different populations are capable of interbreeding. There is regular gene flow in the gene pool.
- In one locality some **physical barrier** isolates a population geographically. This barrier could be a river, mountain range or stretch of ocean. *Gene flow with the ancestral population is prevented.*
- The isolated population and ancestral population are subjected to *different selection pressures*. For example, different selection pressures may arise because of differences in the food source or in the climate (wetter/drier and warmer/cooler). If the isolated population is small (which often happens on islands), genetic drift (random fluctuations in allele frequencies) and the founder effect (the individuals that became isolated were not genetically representative of the ancestral population) also affect its genetic composition.
- The isolated and ancestral populations *diverge genetically* and have, for example, different morphological appearances. If members of the isolated population remain potentially capable of interbreeding with members of the ancestral population, they do not form a different species (though they may be regarded as a subspecies).
- The two populations may *diverge genetically to the extent that interbreeding is prevented*. A new species has evolved. In the speciation of fruit flies (*Drosophila spp.*), genetic differences arose which caused differences in the size of genitalia and mating behaviour.

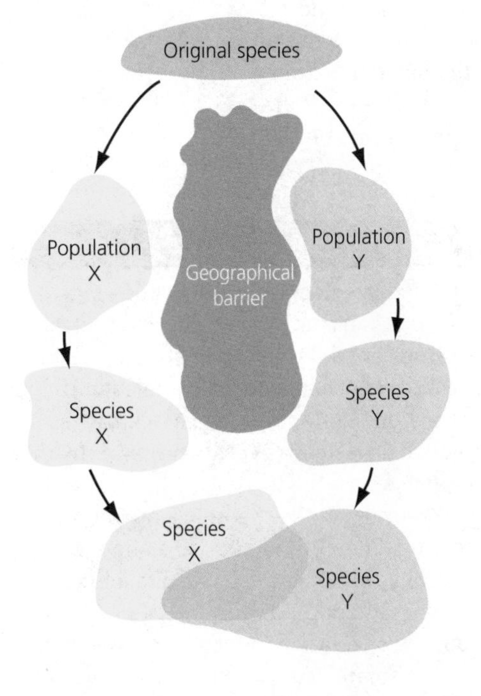

Figure 43 Allopatric speciation

Knowledge check 74

Islands frequently have different, though closely related, species to those on the mainland. Explain why.

If the new species and the ancestral species come together again because the barrier is no longer effective, there will be further interaction. Since they are most likely to have similar niches, there will be interspecific competition with two possible outcomes:

- One species (possibly the new species) will eliminate the other.
- Both species will evolve further and there will be niche divergence (see Figure 44). This is because the most intense competition will occur between the members of both species in the area of niche overlap and so they will be selected against.

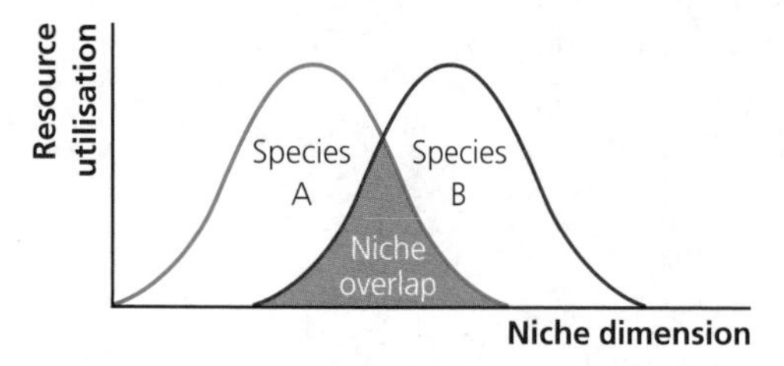

Competition is more intense in the area of niche overlap, resulting in selection against individuals with overlapping characteristics

The niche dimension might represent the 'size of seeds eaten', as in the case of Galapagos finches (where the character in question is beak size), and resource utilisation would be 'amount of seed eaten'

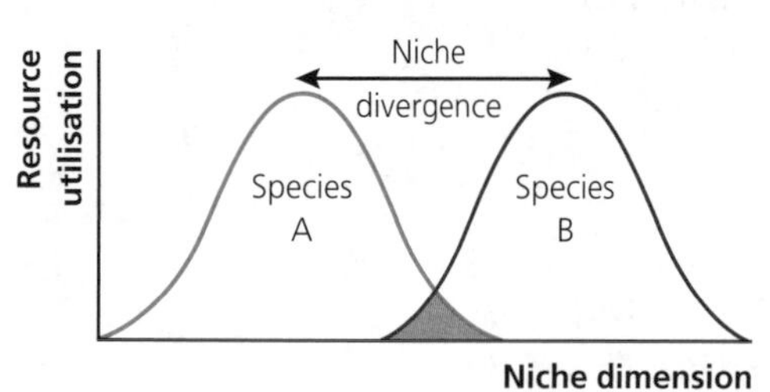

There is further evolutionary change (in the case of Galapagos finches those birds in the range of overlap were selected against) leading to divergence in the niches of the two specie

Figure 44 Niche divergence between two closely related species

The Galapagos finches, observed by Darwin, are a well-known example of speciation and niche divergence. The Galapagos Islands are a group of 19 islands in the Pacific Ocean colonised by a small group of ancestral finches from South America almost 3 million years ago. On the outer islands, different species evolved to give rise to the 13 species of finch. Two species of finch (*Geospiza fortis* and *G. fuliginosa*) illustrate niche divergence: where they are found separately on different islands their beak size is similar (the size of food eaten is similar) but where they are found together on the same island their beak sizes are quite distinct.

Summary

- The gene pool is the sum of all the alleles of all the genes in a population.
- The Hardy–Weinberg equation ($p^2 + 2pq + q^2 = 1$) describes the relationship between allele frequencies (p and q) and genotype frequencies, provided that certain conditions are met. The Hardy–Weinberg principle states that, with these assumptions, the allele and genotype frequencies remain constant. Changes in allele frequency may indicate selection.
- Selection acts on the variation in a population. Genetic variation can be caused by meiosis and fertilisation (which produce new combinations of existing alleles) and mutation (which gives rise to new alleles).
- In selection, the fittest individuals (those with an advantageous allele) survive to produce the most offspring.
- In stabilising selection, mean values of a range for a characteristic are at a selective advantage (extremes are selected against). This maintains the constancy of a characteristic, the adaptive norm.
- In directional selection, one extreme has a selective advantage over the other extreme. The range of values for the population shifts towards the extreme with the selective advantage. Directional selection leads to a change in allele frequency, which is the basis of evolutionary change in the population.
- Allopatric speciation takes place when populations become isolated geographically. In separate areas, different selection pressures operate to cause genetic divergence that is sufficient to prevent interbreeding of the new species.

Kingdom Plantae

All plants are multicellular, composed of eukaryotic cells connected by plasmodesmata through cellulose walls. They are photosynthetic and many cells contain chloroplasts. The kingdom is subdivided into divisions (or phyla) according to the presence or absence of specialised conducting tissue (the vascular tissue) and whether spores or seeds are the means of reproduction and dispersal (Figure 45).

Synoptic links

Plant cell structure, transport and classification

In AS Unit 1 you learned about plant cell structure; and, in AS Unit 2, about transport in plants and the features of plants.

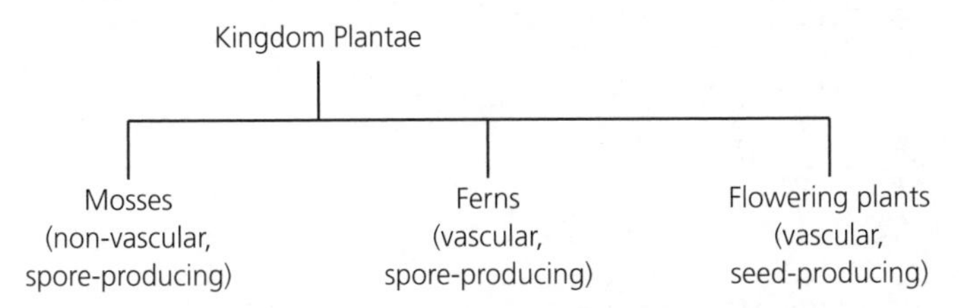

Figure 45 Classification in kingdom Plantae

Knowledge check 75

Many mosses are intolerant of pollution. Suggest why.

Plants have evolved many adaptations for life on land, though these differ in their effectiveness among the different divisions (reflecting stages in the evolution of plants).

Mosses

Mosses, while multicellular plants, are not differentiated into true leaves, stems or roots, though there are analogous structures, e.g. they have 'leaf-like' structures, though these may consist of a single layer of cells (lacking epidermal or mesophyll layers) — see Figure 46. Key features are described in Table 14.

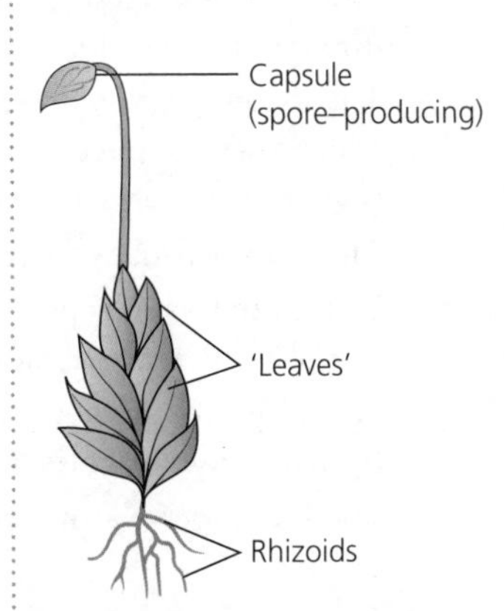

Figure 46 A moss

While mosses are terrestrial plants, they are restricted to moist habitats, where they form the ground layer in many ecosystems.

Table 14 The features of mosses their benefits and restrictions

Feature	Benefit	Restriction
Possess rhizoids (filaments of cells) for anchorage	Can colonise bare rock surfaces, so are often pioneer plants	Distribution is limited to areas with water and ions close to the surface, since rhizoids do not penetrate soils deeply
Absence of cuticle (except in spore-producing capsule) and stomata	Water and ions are obtained directly from the damp environment across the cell walls of all tissues	Are restricted to moist habitats (though can survive periods of dormancy in a dry state)
Absence of vascular tissue	Since water and ions are absorbed by all cells, specialised conduction tissue is not essential	Are only a few centimetres in height; support is by turgor in the cells of the moss
Spores are single-celled haploid reproductive structures	Dispersal of the moss — spores possess a tough wall, enabling them to disperse in air without drying out	Spores germinate in moist conditions only

Ferns

Ferns are well differentiated, with leaves, stems and roots (Figure 47).

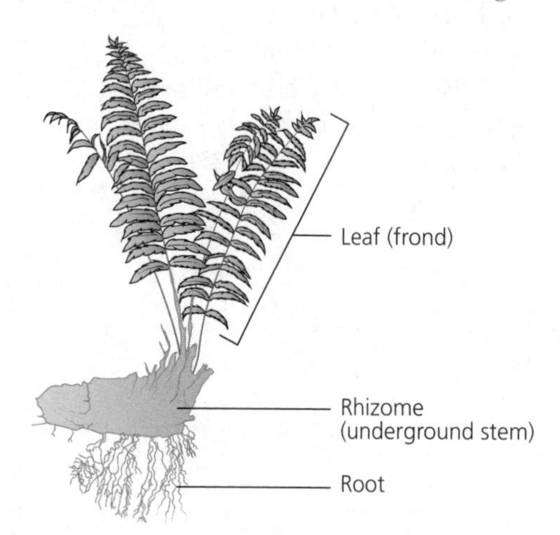

Figure 47 A fern

Key features of ferns include the following:

- The stem is a **rhizome** (horizontal, underground stem) with leaves projecting upwards — since stems are not upright, most ferns are only about 1 m in height (though in tropical rainforest tree ferns are found).
- The leaves (referred to as fronds) possess a **waterproof cuticle** and **stomata** (over which there is fine control of opening and closure) so that transpirational water loss is greatly reduced.
- The roots penetrate the soil for anchorage and for the **absorption of water and ions.**
- **Vascular tissue** is present, with xylem to conduct water and ions from the roots and phloem to distribute photosynthetic products from the leaves.
- While there is support by **turgor** in cells, additional support is provided by **lignified xylem cells** (tracheids) in the vascular tissue.
- The leaves possess numerous spore-producing bodies providing huge reproductive capacity.
- The **spores**, reproductive structures which disperse the species, have a tough outer wall though require water for germination.

Compared with mosses, ferns exhibit a greater adaptation to terrestrial life, though moist conditions are required for reproduction.

Exam tip

Moss and fern spores have tough outer walls containing sporopollenin (also found in pollen grain walls). This chemical is very stable, and its resistance to environmental stresses enables spores to disperse in air without drying out.

Knowledge check 76

Suggest why tree ferns are found mainly in the understorey of tropical rainforest, below the tree canopy.

Flowering plants (angiosperms)

Flowering plants have leaves, stems and roots, and flowers (in which seeds may be produced) — see Figure 48. The stems are upright so that leaves and flowers are held higher above the ground. Flowering plants possess vascular tissue xylem **vessels** for efficient conduction of water and ions, and in some, extra xylem is laid down annually to form wood. Such woody shrubs and trees are more competitive in raising their leaves even higher in the canopy.

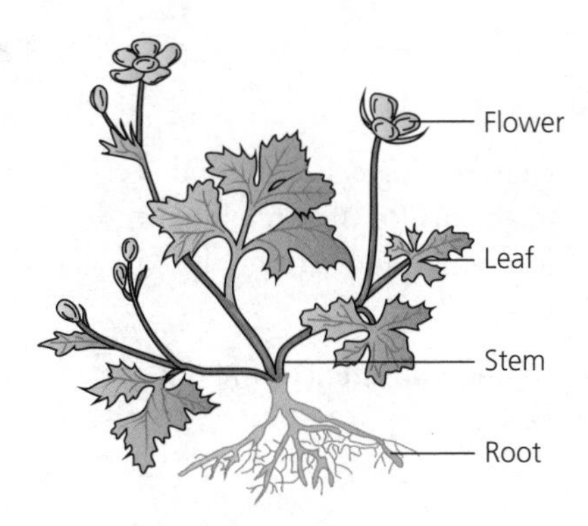

Figure 48 A flowering plant (a buttercup)

The leaves of flowering plants are covered by waxy cuticle and possess stomata (in common with ferns). However, some species that live in dry habitats have extra features to restrict water loss, i.e. xerophytes have adaptations such as rolled leaves and sunken stomata. Furthermore, all flowering plants produce seeds with a tough outer coat providing an ability to withstand desiccation though requiring water for germination.

Synoptic links

Xerophytes

In AS Unit 2 you learned about xerophytic adaptations.

As a result of their range of adaptations, flowering plants (with more than 250 million species) are the dominant plants on Earth today.

Practical work

Study appropriate living and preserved specimens, prepared slides and photographs.

Exam tip

Examiners may ask you to 'compare and contrast' the features of two plant types. 'Compare and contrast' emphasises the need for both similarities and differences — students often ignore the similarities.

Exam tip

Seeds have advantages over spores as agents of dispersal. They contain a food store for germination of the seedlings and a tougher coat better adapted to withstanding desiccation.

Knowledge check 77

How are flowering plants better adapted to a terrestrial existence than ferns?

Summary

- Plants are photosynthetic multicellular eukaryotes. Among plant divisions (phyla), there is an evolutionary trend for the development of adaptations for life on land.
- Mosses possess rhizoids for anchorage though absorption of water and ions takes place over the body surface since there is no epidermis (or cuticle); there is no vascular system and mosses are small, forming the ground layer in moist environments.
- Mosses and ferns disperse via spores, while flowering plants produce seeds.
- Ferns and flowering plants possess vascular tissue, including lignified xylem which provides additional support, a waterproof waxy cuticle and an epidermis with stomata.
- Flowering plants are often the dominant plants in any ecosystem because some are able to produce extra xylem (woody shrubs and trees) so that leaves are carried high into the canopy. Some flowering plants, living in dry habitats, have xerophytic adaptations.

Kingdom Animalia

All animals are multicellular, composed of eukaryotic cells lacking cell walls and chloroplasts. They are heterotrophs, feeding on other organisms (or their remains) and digesting them in a gut cavity, called the **enteron**. Most animals move from place to place in search of food.

The various phyla that make up the kingdom Animalia exhibit differences in body structure, particularly in relation to support and feeding.

Knowledge check 78

What features do all animals have which distinguish them from plants and fungi?

Phylum Cnidaria

The cnidarians are aquatic and include sea anemones, jellyfish, corals and the freshwater *Hydra*. Their body structure (see Figure 49) has the following features:

- The body is **radially symmetrical**, so it can respond to its environment through 360°, i.e. cnidarians can catch prey in any direction.
- The gut cavity (enteron) is sac-like, with only one entrance, a **mouth** (through which food may be ingested and undigested food egested).
- Tentacles contain 'stinging cells' (**cnidicysts**, which give the phylum its name) used to catch food, which is then conveyed by the tentacles to the mouth.
- The body is supported by the surrounding aqueous medium and a **hydrostatic skeleton** formed by the fluid-filled enteron.
- Movement is limited: jellyfish movement is largely dictated by the ocean currents, and while *Hydra* is often sedentary (anchored on a plant or rock), it can undertake cartwheel movements.

Radially symmetrical Situation where body parts are arranged in a circle around an imaginary line through their mouth and gut cavity.

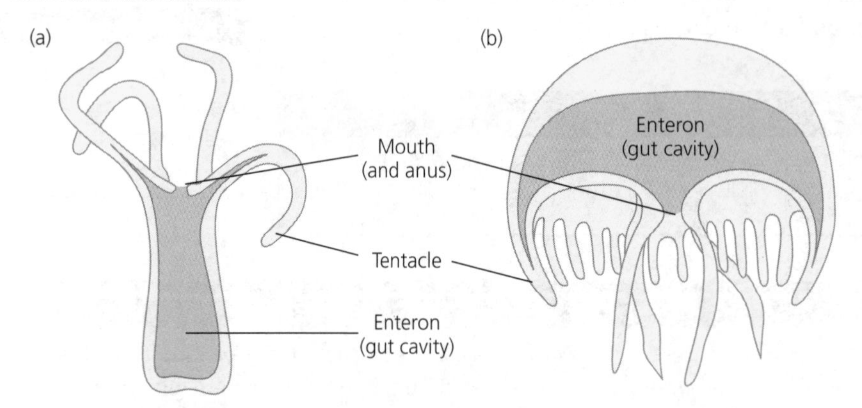

Figure 49 The body structure of (a) *Hydra* and (b) a jellyfish

Phylum Platyhelminthes

The platyhelminths are commonly called flatworms and include the free-living planarians and the parasitic liver fluke. Planarians live in a variety of habitats, including fresh and salt water and moist soil, while the liver fluke lives in animal tissues as parasites.

The body of a planarian (see Figure 50) has the following features:

- It can move forward in search for food, so has a body which is **bilaterally symmetrical**. Movement in one direction means that a front end (head, with

Bilaterally symmetrical Situation where body parts are arranged on either side of an imaginary line through their mouth and gut cavity.

sensory organs such as eyes, and the likely site for a mouth) evolves, as well as a specialised ventral surface (on which the animal travels) with a distinct dorsal surface. Most planarians are detritivores but there are a few active carnivores.

- It is **dorso-ventrally flattened** and, as a result, has a large surface-area-to-volume ratio.
- The gut has only a single opening (through which food may be ingested and undigested food egested), though is **highly branched**, permeating all parts of the body.
- As a result of the branched gut and overall flattened shape, no cell in the body is far from either the gut or the permeable body surface. Therefore, there is a short diffusion path for metabolites and so flatworms lack a circulatory system.
- The body is supported by an aqueous medium, though well-packed cells in the body provide a supportive role (they are soft-bodied as there is no specialised skeletal system).

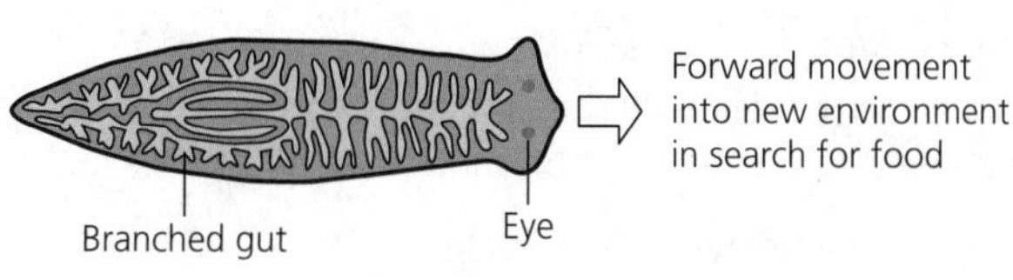

Figure 50 A planarian

Knowledge check 79

Briefly state how the body plan of platyhelminths differs from that of cnidarians.

Knowledge check 80

Distinguish between radial symmetry and bilateral symmetry.

Knowledge check 81

Suggest why flatworms, such as *Planaria*, are poorly adapted for life on land.

Phylum Annelida

The annelids are segmented worms and include the earthworm and lugworm. They inhabit aquatic (marine and freshwater) and moist terrestrial (soil) ecosystems.

The body of an earthworm (see Figure 51) has the following features:

- The body is bilaterally symmetrical, though **round in transverse section**.
- There is a **through (one-way) gut**, with both a mouth (for ingestion) and an anus (for egestion). Earthworms are detritivores.
- The one-way gut allows **regional specialisation**: food is drawn into a muscular pharynx, passed along an oesophagus to be stored in the crop, slowly released into the muscular gizzard where it is crushed, then digested and absorbed in the intestine; this allows the ingestion of food to continue, while previously ingested food is digested.
- Gas exchange takes place over a moist body surface, though a **blood circulatory system** is required for the distribution of gases and food molecules.
- It is **metamerically segmented** (i.e. the body is divided up into many repeating segments, each containing identical copies of nerve ganglia and muscle blocks).
- It is supported by a hydrostatic skeleton provided by segmented, fluid-filled body cavities.
- Locomotion relies on the action of muscles acting on the hydrostatic skeleton, aided by external bristles of chitin, called **chaetae**. For ease of movement, earthworms are long and thin.

Exam tip

The metameric segmentation of muscle blocks allows them to contract independently of each other and yet in a coordinated way, making locomotion more efficient.

Knowledge check 82

Suggest why earthworms do not require special organs for the exchange of respiratory gases.

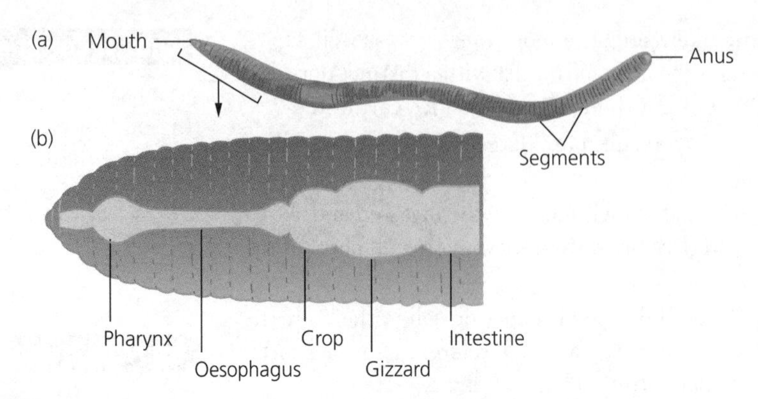

Figure 51 (a) An earthworm, (b) front end ('head') showing regional specialisation of gut

Phylum Arthropoda

The arthropods (crustaceans, centipedes, millipedes, arachnids and insects) live in a wide variety of habitats, both aquatic and terrestrial, including the air. Three-quarters of all animal species are arthropods and insects alone account for more than half the total.

The body of an insect (see Figure 52) has the following features:

- The body is bilaterally symmetrical.
- There is a one-way gut, with both mouth and anus, allowing regional specialisation.
- It is metamerically segmented and has different regions with a fixed number of segments in each region. Insects have a head (six fused segments), thorax (three segments) and abdomen (eleven segments).
- A rigid **exoskeleton of chitin** over the body provides protection, a degree of water retention, and support. Muscles governing movement are attached internally to the exoskeleton of the **jointed limbs** and to **wings for flight**.
- The presence of an exoskeleton limits the size of arthropods (since it needs to be shed prior to a growth spurt) and creates a need for specialised gas exchange surfaces.
- The exoskeleton of insects is covered in a waxy water-resistant cuticle, a terrestrial adaptation, to further reduce desiccation. Crustaceans lack this as they are mostly aquatic.

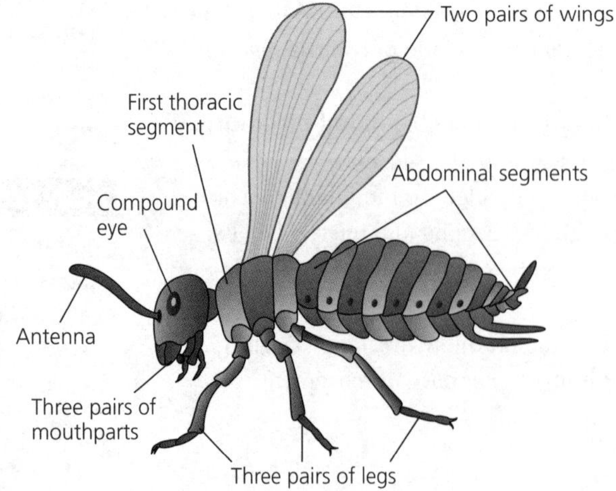

Figure 52 A generalised insect

Phylum Chordata

The chordates include the vertebrates — fish, amphibians, reptiles, birds and mammals.

The body of a mammal has the following features:

- It is bilaterally symmetrical, is metamerically segmented, and possesses a one-way gut with mouth and anus and showing regional specialisation.
- As in all vertebrates, there is a vertebral (spinal) column.
- An **endoskeleton** consists of an internal jointed system of calcified bones to which muscles for locomotion attach externally.

Exam tip

Metameric segmentation in humans is most easily recognised in the separate vertebrae of the vertebral column and, associated with each vertebra, the paired spinal nerves arising from the spinal cord. In some, metameric segmentation can be seen in the blocks of muscle in the abdominal wall constituting the 'six-pack'.

Evolutionary trends in the kingdom Animalia

In the animal kingdom there are the following evolutionary trends:

- from radial symmetry, which suits sedentary organisms, to bilateral symmetry, which is a consequence of directional movement (also producing an anterior and posterior end, and a ventral and dorsal surface)
- from a sac-like gut cavity (enteron) to a through gut, providing the conditions for more efficient feeding and digestion, with one-way processing of ingested food
- towards the development of an internal blood transport system, the inclusion of a rigid skeleton and the possession of an impermeable outer covering, which is better adapted to a terrestrial existence

Knowledge check 83

Explain why some animals show radial symmetry, while others have bilateral symmetry.

Practical work

Study appropriate living and preserved specimens, prepared slides and photographs.

Summary

- Animals are multicellular eukaryotes with cells lacking cell walls. They are heterotrophic, typically possessing a gut cavity (enteron), and most are motile.
- Cnidarians have a radially symmetrical body form and a sac-like enteron which is fluid-filled, forming a hydrostatic skeleton.
- Platyhelminths (flatworms) are bilaterally symmetrical and food is drawn into a blind-ending branched gut. Flatness increases the surface-area-to-volume ratio for gaseous exchange and negates the need for a blood transport system.
- Annelids (metamerically segmented worms), arthropods and chordates are bilaterally symmetrical, and have a one-way gut (with mouth and anus), which allows regional specialisation, and a blood transport system.
- Annelids have a hydrostatic skeleton, while a rigid skeleton is found in arthropods (exoskeleton) and chordates (endoskeleton). Gas exchange takes place on the moist body surface of annelids, but in specialised gas exchange surfaces in arthropods and chordates.

Questions & Answers

The examination

The A2 Unit 2 examination contributes 24% to the final A-level outcome. The paper lasts 2 hours 15 minutes and is worth 100 marks. In Section A (82 marks) all the questions are structured, though with a variety of styles. In Section B (18 marks) there is a single question, possibly with several parts, to be answered in continuous prose.

In the A2 Unit 2 paper the approximate marks allocated for each assessment objective (AO) are as follows:

AO1 Knowledge and understanding: 31

AO2 Application of knowledge and understanding: 43

AO3 Analysis, interpretation and evaluation of scientific information, ideas and evidence: 26

Notice the prominence given to the application of your understanding (AO2).

Skills assessed in questions

While some questions will assess straightforward knowledge and understanding (AO1), most will require you to *apply* your understanding to unfamiliar situations (AO2), and there will be some that ask you to evaluate experimental and investigative work (AO3).

A2 units include some synoptic assessment, involving links between different areas of biology and including concepts from the AS units.

A variety of mathematical skills, including statistical skills, will be tested in some questions while quality of written communication is assessed in Section B.

About this section

This section consists of questions covering the range of topics in the Content Guidance. Answers are provided by two students of differing ability. Student A consistently performs at grade A/B standard. Student B makes a lot of mistakes — ones that examiners often encounter — and grades vary between C/D and E/U.

Each question is followed by a brief analysis of what to look out for when answering the question (shown by the icon ⓔ). All student responses are followed by comments (preceded by the icon ⓔ). They provide the correct answers and indicate where difficulties for the student occurred, including lack of detail, lack of clarity, misconceptions, irrelevance, poor reading of questions and mistaken meanings of examination terms.

Read the question carefully and think about how you will answer before you begin to write. You must ensure that your answer responds appropriately to the command term used, for example 'describe' or 'explain'. In questions set in an unfamiliar context, you may be presented with information that you need to use in constructing a full answer.

Section A Structured questions

Respiration

Question 1 Anaerobic respiration

The flow diagram below shows four steps (labelled 1 to 4) in a metabolic pathway occurring in mammalian skeletal muscle tissue.

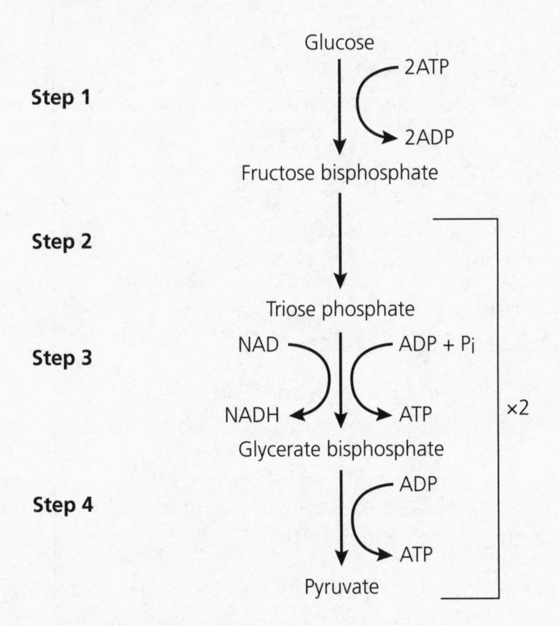

(a) (i) Name the metabolic pathway shown. (1 mark)

(ii) State the net yield of ATP from one molecule of glucose metabolised by this pathway. (1 mark)

(iii) State where in muscle tissue this pathway occurs. (1 mark)

(iv) Explain why glucose is not normally limiting in muscle tissue. (1 mark)

(b) NAD occurs in very low concentration in muscle tissue and needs to be recycled for step 3 to continue. Explain how hydrogen is removed from NADH to recycle NAD when:

(i) oxygen is available (and not limiting) (2 marks)

(ii) oxygen is not available (or is limited) (2 marks)

(c) (i) Explain the advantage of regenerating NAD in mammalian skeletal tissue when the supply of oxygen is limiting. (2 marks)

(ii) Following strenuous exercise in mammals, there is a period when extra oxygen is consumed to repay an oxygen debt. Describe the metabolic changes that occur during this period of excess post-exercise oxygen consumption. (2 marks)

(d) Yeast cells are also able to regenerate NAD when the supply of oxygen is limiting. Describe the metabolic pathway by which NAD is regenerated in these conditions and explain the advantage to yeast. (3 marks)

Total: 15 marks

ⓔ This question depends on your recall of facts and concepts. It should be straightforward as long as you have learned and can remember this part of the course.

Student A

(a) (i) Glycolysis ✓

(ii) 2 ATP ✓

(iii) Sarcoplasm ✓

(iv) The liver contains a store of glucose in the form of glycogen. ✓ a

(b) (i) Hydrogen from NADH passes into the respiratory chain ✓ in the mitochondria and ultimately oxygen is reduced to water. ✓

(ii) Pyruvate acts as a hydrogen acceptor and is reduced ✓ to lactate or ethanol ✓ regenerating NAD for glycolysis to continue. a

(c) (i) Glycolysis can continue. Though only 2 ATPs per glucose, this allows extra ATP to be produced, ✓ over and above that produced aerobically. ✓

(ii) The extra oxygen intake after strenuous exercise is used to respire some of the lactate, ✓ generating extra ATP to reconvert the remaining lactate to glucose ✓ and subsequently glycogen in the liver. a

(d) The pyruvate is decarboxylised, releasing CO_2, ✓ and reduced by NADH to ethanol, ✓ so that NAD is available for glycolysis to continue producing some ATP. This allows yeast to survive for a limited period of time in anaerobic conditions. ✓ a

ⓔ **15/15 marks awarded** a All correct, for full marks.

Student B

(a) (i) Anaerobic respiration ✗ a

(ii) 4 ATP per glucose ✗ a

(iii) Cytoplasm ✓ a

(iv) There are large stores of glycogen in the liver. ✓ a

(b) (i) Hydrogen is passed from NADH to the electron transport chain ✓ where oxygen is the final electron acceptor. ✓ This happens in the mitochondria. b

(ii) There is no hydrogen acceptor and so NAD is not regenerated and the Krebs cycle stops. ✗ c

(c) (i) The regenerated NAD allows step 3 to continue so that step 4 produces ATP. ✓ d

(ii) It goes to the liver where it is converted back to pyruvate and respired aerobically ✓ using the extra oxygen intake. e

(d) Pyruvate is metabolised anaerobically in the yeast, producing CO_2 ✓ and ethanol. ✓ This regenerates NAD and allows ATP production to continue. f

e 8/15 marks awarded a Parts (i) and (ii) are incorrect: anaerobic respiration, though the title of the question, has additional reactions beyond glycolysis, while in (ii) the production of 4 ATPs is the gross yield per glucose. Parts (iii) and (iv) are both correct. b This is sufficient for 2 marks. c This is what happens in the mitochondria if oxygen is not available, but is irrelevant to what happens in the cytoplasm. The regeneration of NAD occurs via the reduction of pyruvate to lactate. No marks awarded. d This is fine for 1 mark, but Student B should have noted that this is extra ATP over and above that produced aerobically. Aerobic respiration is occurring at a maximum since there is a greater supply of oxygen to the muscles through increased breathing and heart rates during exercise. e Again this is only sufficient for 1 mark. Extra ATP is produced from the metabolism of some of the lactate and this is used to convert most of the lactate back to glucose. f Student B understands the anaerobic pathway in yeast and is awarded 2 marks. However, for the third mark, it should have been clear that this is a completely different strategy than for muscle. It allows yeast to survive for a period in anaerobic conditions — muscle would die in anaerobic conditions.

Question 2 Aerobic respiration

The figure below represents the aerobic respiration of glucose and fatty acid.

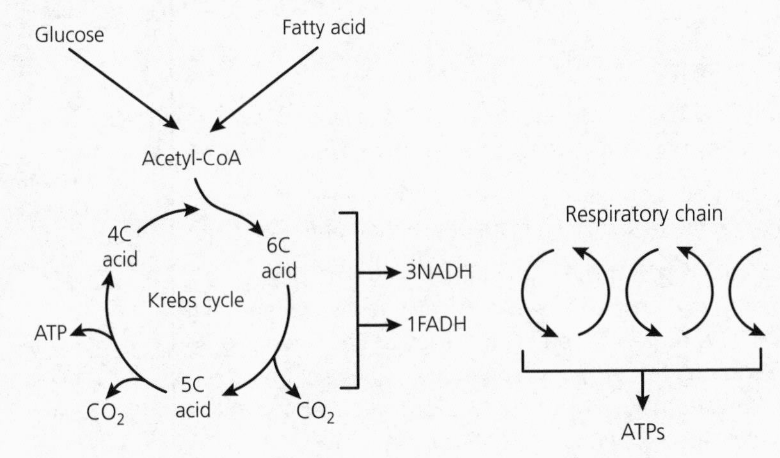

(a) State the precise location of each of the following:

(i) the Krebs cycle (1 mark)

(ii) the respiratory chain (1 mark)

(b) (i) Explain how the Krebs cycle generates the reduced coenzymes NADH and FADH. (1 mark)

(ii) The figure shows one molecule of ATP produced by substrate-level phosphorylation. Explain the term substrate-level phosphorylation. (1 mark)

(c) (i) Explain how the respiratory chain results in the synthesis of ATP from the reduced coenzymes NADH and FADH. (4 marks)

(ii) Describe the role of oxygen in the operation of the respiratory chain. (1 mark)

(iii) What term is used to describe the synthesis of ATP from the operation of the respiratory chain? (1 mark)

(d) Linoleic acid is an essential fatty acid, occurring in many plant seeds, required for the biosynthesis of prostaglandins and cell membranes, though it may also be used as a respiratory substrate. The equation for the respiration of linoleic acid is:

$$C_{18}H_{32}O_2 + 25O_2 \rightarrow 18CO_2 + 16H_2O$$

(i) Calculate the respiratory quotient (RQ) for the aerobic respiration of linoleic acid. Show your working. (2 marks)

(ii) Suggest the advantages of plant seeds using lipids as energy storage molecules. (2 marks)

Total: 14 marks

e In this question it should be relatively easy to gain marks as long as you are well prepared and answer precisely. Notice that part (c) (i) is worth 4 marks, so you will need to make sure that you provide a detailed answer. In part (d) (i), show your working so that you have a chance of getting a mark even if you make a mistake and arrive at an incorrect answer.

Student A

(a) (i) The matrix of the mitochondria ✓

(ii) The cristae in the mitochondria ✓ a

(b) (i) The Krebs cycle involves many dehydrogenase reactions so that hydrogen is made available to be picked up by NAD and FAD. ✓

(ii) Substrate-level phosphorylation takes place when the conversion of one substrate to another yields energy for the production of ATP. ✓ a

(c) (i) Hydrogen is carried from NADH through the initial carriers, flavoprotein and coenzyme Q, while from FADH hydrogen is transferred directly to coenzyme Q. ✓ Thereafter electrons are transferred through cytochromes ✓ and protons released into the matrix. These carriers are at progressively lower energy levels ✓ and at specific steps sufficient energy is available to synthesise ATP molecules. ✓

(ii) Oxygen is the final electron acceptor and is reduced to water, having combined with electrons from the respiratory chain and free protons. ✓

(iii) Chemiosmosis is the process by which ATP is coupled to electron transport. ✓ a

(d) (i) $RQ = \frac{\text{amount of } CO_2 \text{ produced}}{\text{amount of } O_2 \text{ consumed}}$ ✓ $= \frac{18}{25} = 0.72$ ✓

(ii) Lipids have more than twice the energy per gram as carbohydrates. ✓ This is more weight efficient, ✓ an important factor if seeds are to be dispersed away from the parent plant to limit competition and to colonise new habitats. e

e 14/14 marks awarded a Excellent answers throughout. The answer provided in (c) (iii), chemiosmosis, is not part of the specification but since it is correct the mark is awarded.

Student B

(a) (i) Mitochondrial matrix ✓ a

(ii) Inner mitochondrial membranes ✓ a

(b) (i) Hydrogen is released during the reactions of the Krebs cycle. ✗ b

(ii) This is the synthesis of ATP which does not involve the ETC. ✗ c

(c) (i) Electrons pass from carriers at a high energy level to carriers at a lower energy level. ✓ At particular points in the chain the energy available is enough for an ATP to be produced. ✓ d

(ii) The electrons from the ETC, along with hydrogen ions in the matrix, are combined with oxygen, producing water. ✓ e

(iii) Oxidative phosphorylation ✓ e

(d) (i) 1.39 ✗ f

(ii) There is more energy in a gram of lipid than in a gram of carbohydrate. ✓ g

e 7/14 marks awarded a Both correct for 2 marks. b This lacks sufficient detail — the student should say that the release of hydrogen is due to the action of dehydrogenase enzymes. No mark gained. c It is not sufficient to say what substrate-level phosphorylation is not. No mark given. d This lacks detail since there is no mention of the carriers involved or that, while initially hydrogen is transferred, thereafter it is their electrons. Still, there are two correct points for 2 marks. e Both parts correct for 2 marks. f This gains no marks — the answer is wrong and no working is shown, though it appears that the student has used the formula the wrong way round. g There is no mention that carrying lipid means that the seed can be lighter than if storing carbohydrate. 1 mark awarded.

Photosynthesis

Question 3 Photosynthesis

(a) The electron micrograph shows a section through a chloroplast.

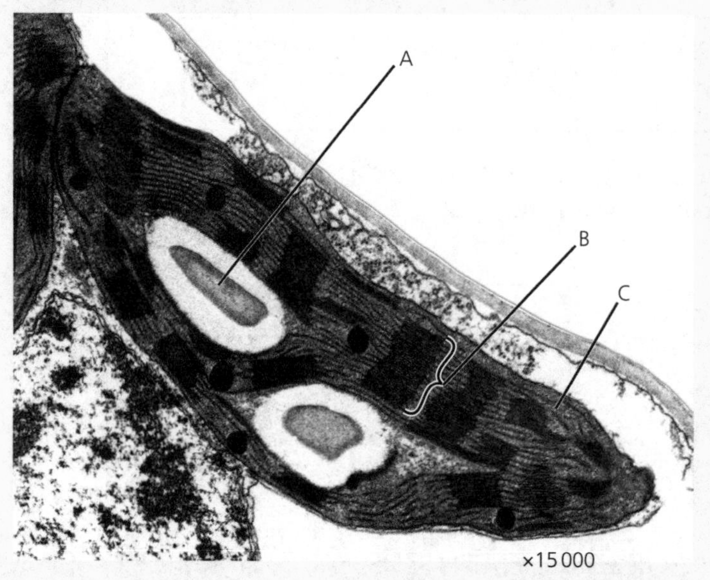

Name the structures labelled A to C. (3 marks)

(b) The diagram below shows the light-dependent reactions of photosynthesis on a thylakoid membrane.

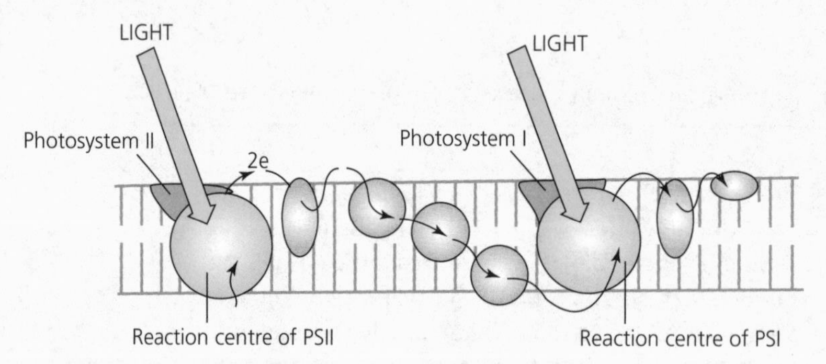

(i) The photosystems consist of different pigments, including chlorophyll *a*, chlorophyll *b* and carotenoids. Explain the advantage of the photosystems possessing different pigments. (2 marks)

(ii) Explain what happens in a photosystem to cause an electron to be emitted from its reaction centre. (2 marks)

(iii) Each photosystem is able to replace the electrons lost due to photoactivation. Explain the process by which photosystem II receives replacement electrons. (2 marks)

(iv) **Atrazine, used as a weedkiller, inhibits the action of photosystem II. Plant death results from starvation. Explain how inhibition of photosystem II would prevent the synthesis of carbohydrates, such as glucose.** (3 marks)

(c) **In an investigation of the light-independent reactions, illuminated photosynthesising tissue, supplied with carbon dioxide for some time, was then deprived of carbon dioxide. The relative amounts of glycerate phosphate (GP) and ribulose bisphosphate (RuBP) during the investigation are shown in the graph below.**

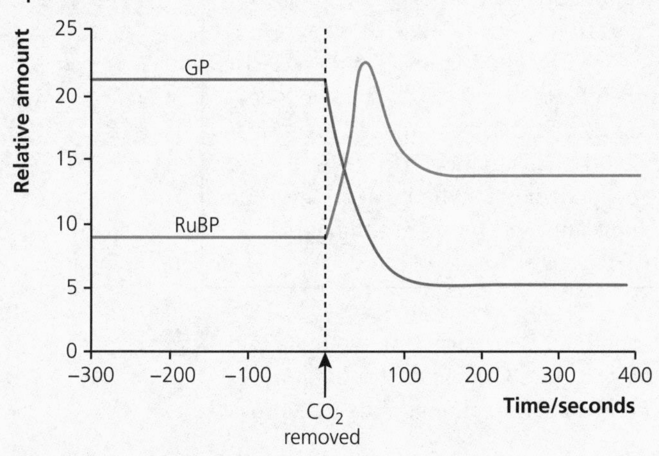

Suggest explanations for the changes in the amounts of GP and RuBP. (5 marks)

Total: 17 marks

ⓔ This question includes all three assessment objectives. Part (a) asks you to identify components of a chloroplast EM (AO1). Part (b), about the light-dependent reactions, assesses AO2 since you are unlikely to have seen this stage of photosynthesis illustrated in this way before. Don't panic if you see something unfamiliar and make sure that you read the questions carefully. In part (c), you are asked to analyse and interpret the results of an investigation (AO3) showing changes in intermediates of the light-independent reactions. Making a quick sketch of the Calvin cycle should help you with this. Remember that the level of an intermediate will depend on its rate of production *and* usage.

Student A

(a) A = starch grain ✓ B = thylakoid ✗ C = stroma ✓ a

(b) **(i)** Each pigment absorbs at a different wavelength, ✓ which increases the range of wavelengths ✓ utilised in photosynthesis.

(ii) As the pigment molecules absorb light, electrons in them become excited. ✓ This energy is passed on by resonance to the reaction centre from which a high-energy electron is emitted. ✓

(iii) Hydroxyl (OH^-) ions, from the dissociation of water, supply PS II with electrons ✓ after which OH radicals form oxygen and water. ✓

(iv) The inhibition of PS II will mean that neither ATP ✓ nor NADPH will be produced. ✓ Both are required for the production of triose phosphate ✓ from which carbohydrate is formed. b

(c) When CO_2 is available, RuBP is converted into GP catalysed by the enzyme rubisco. When there is no CO_2, RuBP cannot be turned into GP so its level increases. ✓ Since GP is not being produced its level falls. ✓ GP also falls because it is still being turned into triose phosphate, ✓ as ATP and NADPH are still being produced in the light-dependent reactions. ✓ After the level of RuBP rises it falls again, possibly because RuBP is being used in other reactions. ✓ The amount of RuBP levels off and is higher than it was before CO_2 became unavailable, presumably because it is still being recycled from triose phosphate. (✓) The recycling requires ATP (✓) but the light-dependent reactions are still working. c

e 16/17 marks awarded a A and C are correct, but B is a granum — a thylakoid is just one element of a granum. 2 marks gained. b Full answers throughout. c Comprehensive answer, well thought out and clearly explained. The maximum 5 marks are awarded.

Student B

(a) A = starch grain ✓ B = granum ✓ C = matrix ✗ a

(b) (i) The different pigments have a different absorption spectrum. ✓ b

(ii) The energy level of electrons is raised by light absorption. ✓ c

(iii) The photolysis of water provides an electron to PS II. ✓ d

(iv) If Atrazine inhibits PS II then no ATP will be produced ✓ and so there will be no carbohydrate. e

(c) RuBP combines with CO_2 to form GP. So when CO_2 is no longer there, the Calvin cycle gets stuck at RuBP and the amount rises, ✓ while GP is no longer produced and so its amount falls. ✓ f

e 8/17 marks awarded a C is the stroma, the special name for the matrix in the chloroplast. A and B are correct for 2 marks. b There is no mention of a greater overall usage of light energy. 1 mark only. c The student fails to say that excited electrons collectively resonate energy through to the reaction centre. 1 mark scored. d This lacks the detail that, in consequence, O_2 is produced. 1 mark gained. e Again this is incomplete. If there is no electron flow from PS II then NADPH will not be produced; and both ATP and NADPH are required for the production of triose phosphate. 1 mark only. f Student B has two obvious points but fails to elaborate or make further suggestions about, for example, the decrease in RuBP after an initial increase. 2 marks awarded.

DNA as the genetic code

Question 4 Polypeptide synthesis

(a) Polypeptide synthesis involves two stages: the transcription of the DNA into mRNA and the translation of mRNA to produce a polypeptide. Since the DNA is made up of exons and introns, the transcription stage makes pre-mRNA.

(i) Name the enzyme involved in transcription. (1 mark)

(ii) Describe how post-transcriptional pre-mRNA is modified before it is used in the synthesis of a polypeptide. (2 marks)

(iii) Explain how the modification of the pre-mRNA allows several different forms of a polypeptide to be coded for by just one gene. (2 marks)

(b) The figure below shows a phase in the translation of functional mRNA.

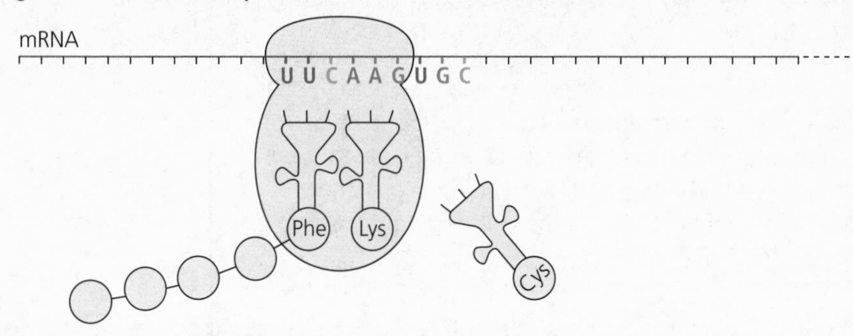

Describe, precisely, what happens at the next phase in the synthesis of the polypeptide. (5 marks)

(c) The table below shows part of the genetic code, the mRNA codons, for some amino acids.

		Second base					
		U	C	A	G		
First base	U	Phenylaline Leucine	Serine Serine	Tyrosine –	Cysteine Tryptophan	C G	Third base
	A	Ileucine Methione	Threonine Threonine	Asparagine Lysine	Serine Arginine	C G	

The codon sequences, shown below, are those for part of the original functional mRNA and those of three different mutations.

Original mRNA ...A C C U U G A U G...

Mutation 1 ...A C G U U G A U G...

Mutation 2 ...A A C U U G A U G...

Mutation 3 ...A C U U G A U G (C)...

(i) Determine the amino acid sequence translated by the original mRNA. (1 mark)

(ii) Explain how each mutation may affect the polypeptide translated by each altered code. (6 marks)

Total: 17 marks

ⓔ Parts (a) and (b) are relatively straightforward, involving recall of the transcription and translation of DNA as the genetic code (AO1), though notice that there are 5 marks for (b) so your answer must be sufficiently detailed. Part (c) requires you to use the genetic code (AO2) to determine the amino sequence and, in the second subpart, to deduce the effects of three mutations — again notice that for each mutation 2 marks are available for a full answer.

Student A

(a) (i) RNA polymerase ✓ a

(ii) After pre-RNA has been made, the introns are separated from the exons. ✓ The exons are then spliced together to make a continuous chain. ✓ This is the functional mRNA. a

(iii) The exons can be spliced together in different ways, ✓ resulting in different mRNAs, each coding for a different amino acid sequence. ✓a

(b) Once the amino acid at the P-site forms a peptide bond with the amino acid at the A-site (Phe to Val) ✓ its tRNA is released into the cytoplasm to pick up another amino acid. ✓ The ribosome moves along one codon length so that the tRNA–amino acid at the A-site is transferred to the P-site ✓ and the A-site becomes available for another tRNA ✓ carrying a specific amino acid ✓ to move in according to the base-pairing relationship between codon and anticodon. (✓) a

(c) (i) Threonine, cysteine, methionine ✗ a

(ii) Mutation 1 causes no change to the sequence of amino acids ✓ since the same amino acid is coded for ✓ — the code is degenerate. In mutation 2, threonine is replaced by asparagine ✓ so that there may well be a change in the polypeptide as the bonds holding the tertiary structure may be different. ✓ Mutation 3 shows a base deletion resulting in a frameshift so that all the amino acids after the deletion are changed; ✓ the polypeptide structure will be different and the protein will be non-functional. ✓ a

ⓔ **16/17 marks awarded** a Correct answers throughout, except in part (c) (i) where cysteine is wrong since its codon is UUC — obviously a slip, with the student misreading C for G. In (b) the maximum of 5 marks is awarded (though this very complete answer shows six correct responses).

Student B

(a) (i) Polymerase ✗ a

(ii) The exons and introns are separated ✓ and the introns joined up with one another. ✗ b

(iii) It is possible for the introns to be joined together in different ways ✓ so that different polypeptides are made. c

(b) Once the tRNA in the P-site has formed a peptide bond with its amino acid, ✗ the tRNA goes off into the cytoplasm to pick up another amino acid. ✓

The tRNA in the A-site moves to the P-site, ✓ leaving the A-site empty for the next tRNA with its amino acid to move in. ✓ d

(c) (i) Threonine, leucine, methionine ✓ e

(ii) Mutation 1 — though the third base is changed, threonine is still coded for. ✓ Mutation 2 — threonine is replaced by asparagine. ✓ Mutation 3 — the third base is deleted, changing all the amino acids. ✓ f

e **9/17 marks awarded** a No mark since there are two polymerase enzymes (DNA and RNA). b The first point is just enough for 1 mark, but in the second point Student B has confused introns and exons. c Having already confused introns with exons, there is no further penalty since the mark is for alternative splicing. The second point repeats the question, so there is no mark for that. 1 mark gained. d Peptide bonds form between neighbouring amino acids, not between an amino acid and its tRNA. The student gains 3 marks. e Correct for 1 mark. f The student understands the genetic code alteration in each case, but fails to explain the effect of each mutation in altering a polypeptide's tertiary structure and functionality. 3 marks are scored.

Question 5 Epigenetics and microarray technology

(a) Studies have shown that children born during the period of the Dutch famine of 1944–1945 had increased rates of coronary heart disease and obesity after maternal exposure to famine during early pregnancy, compared with those not exposed to famine.

(i) What other factors would need to be considered in the control group, apart from 'not being exposed to famine'? (2 marks)

(ii) It was considered that 'exposure to famine during early pregnancy' was linked to epigenetic changes in the children born. Explain the term 'epigenetics'. (2 marks)

(iii) Exposure to famine during pregnancy was found to be associated with less methylation of the growth factor gene (*IGF2*). Explain the consequence of reduced methylation of this gene. (2 marks)

(iv) Epigenetic change can also be caused by histone acetylation. State the effect of histone acetylation. (2 marks)

(b) Epigenetic changes occur during ageing. Occasionally this might lead to the activation of a cancer-causing gene (oncogene). Explain how a DNA microarray (DNA chip) might be used to determine whether an oncogene is being expressed in, say, breast tissue. (5 marks)

Total: 13 marks

e Part (a) assesses AO3 in (i) — control of variables to allow conclusions to be valid — and AO1 in the rest — your understanding of epigenetic changes — though (iii) involves a deduction about the effect of the gene and so is AO2. Part (b) tests your understanding of microarray technology in assessing gene expression, and is quite complicated. You will need to include enough detail to have a chance of getting all 5 marks — it may be helpful to jot down the sequence of events before answering the question.

Student A

(a) **(i)** The pregnant women in the control group should be from the same ethnic group ✓ and of a similar age range. ✓

(ii) Epigenetics involves changes in gene expression due to changes in the cell environment, ✓ and not due to changes in the genetic code. ✓

(iii) Reduced methylation will cause a gene to be activated. ✓ Increased activity of the gene will cause increased growth, affecting obesity or putting a strain on the heart. ✓

(iv) Histone acetylation causes DNA to be less tightly wound around the histones so that DNA is more readily transcribed. ✓ The gene is activated. ✓ a

(b) Suspected cancerous breast tissue and healthy tissue are excised and mRNA extracted from each. ✓ Only genes which are switched on will produce mRNA. Single-stranded cDNA is produced from mRNA using reverse transcriptase, ✓ and tagged with a coloured fluorescent chemical — healthy tissue cDNA is tagged red and breast tissue cDNA tagged green. ✓ A mixture of cDNA from both tissues is added to an appropriate microarray and the tagged cDNA will hybridise with any complementary probe. ✓ If a laser scan of the array shows spots signalling green, these indicate that genes are switched on in the breast tissue only. ✓ If the probe in the array for the oncogene shows green and not red, then the oncogene is being expressed in the breast tissue and not in the healthy tissue. (✓) b

e **13/13 marks awarded** a Correct and full answers throughout. b Student A has 6 correct points and receives the maximum mark of 5.

Student B

(a) **(i)** Same genetic background. ✓ Same age. ✓ a

(ii) Epigenetics is the study of chemicals inside the cell which can cause genes to be activated or silenced. ✓ b

(iii) The gene is silenced. ✗ c

(iv) Histone acetylation will activate genes. ✓ d

(b) DNA is extracted, ✗ made single-stranded and labelled with a fluorescent dye. This is done for both diseased and healthy tissue. ✓ The labelled DNA is added to the DNA chip and joins with the complementary probes in particular spots. ✓ Laser scanning and analysis of the results will indicate which genes are activated and which inactivated. e

e 6/13 marks awarded a Both points correct for 2 marks. b This is fine for 1 mark since the student says that the effect is due to chemicals in the cell (there are regulatory genes which switch other genes on or off), but the student fails to note that the base sequence (genetic code) remains unchanged. c Incorrect — the question asked about the effect of *reduced* methylation. d This is correct for 1 mark, but there is no mention about how acetylation of histones will bring about this effect. e There are mistakes and a lack of detail here: DNA is not extracted, it is the mRNA which is used to determine gene expression (there may be confusion with genotyping in which DNA is analysed); further, the student has provided no detail about how the scanned microarray is analysed. Nevertheless, Student B notes that genetic material from both experimental and control tissue incorporates fluorescent dye, and that DNA joins with complementary probes on the microarray, for 2 marks.

Gene technology

Question 6 Transgenic and gene-silenced tomato plants

(a) The figure below shows the use of the Ti plasmid, from the bacterium *Agrobacterium tumefaciens*, to transfer a gene, the *Bt* gene, into tomato plants. The *Bt* gene, obtained from another bacterium, *Bacillus thuringiensis*, codes for the production of Cry protein which is toxic to insect larvae such as caterpillars.

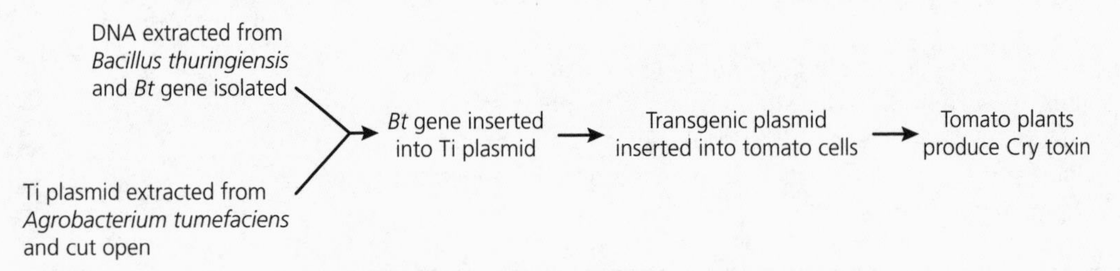

(i) Describe how the *Bt* gene would be isolated from the bacterium *Bacillus thuringiensis*. (2 marks)

(ii) Describe how the isolated *Bt* gene would be transferred into the Ti plasmid. (2 marks)

(iii) Describe how the transgenic plasmid would be inserted into tomato cells. (2 marks)

(iv) Explain why it is advantageous to have tomato plants that are genetically modified to produce Cry toxin. (2 marks)

(v) Suggest two possible problems that may be associated with growing these genetically modified plants. (2 marks)

(b) **When a tomato ripens, it becomes softer because of the action of an enzyme called polygalacturonase (PG), which causes the cells to separate. The Flavr Savr tomato is a genetically modified tomato that contains an antisense gene. The antisense gene is a mirror image of the *PG* gene and transcribes mRNA that is complementary to the mRNA from the *PG* gene. This anti-mRNA combines with, and inactivates, the *PG* mRNA. The process is summarised in the figure below.**

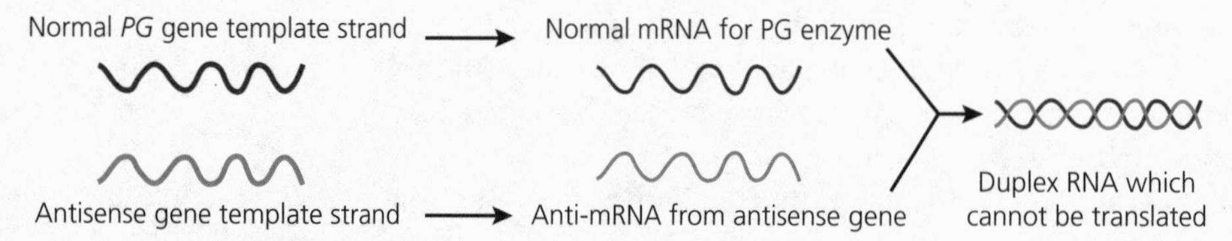

(i) **The base sequence of part of the template strand of the *PG* gene is ATGCATC. Determine the base sequence on the corresponding part of the:**
- **normal mRNA (for the PG enzyme)**
- **anti-mRNA (transcribed from the antisense gene)** (2 marks)

(ii) **State two ways in which the structure of the RNA duplex differs from the structure of DNA.** (2 marks)

(iii) **Explain why the RNA duplex cannot be translated to synthesise the enzyme PG.** (2 marks)

(iv) **Suggest an advantage of reducing PG activity during tomato ripening.** (2 marks)

Total: 18 marks

ⓔ Parts (a) and (b) ask questions about different types of transgenic tomatoes. Both involve straightforward understanding (AO1), application of understanding (AO2) and an evaluation of social issues (AO3). Make sure that you read the information provided and study the figures before answering the questions, and ensure that questions are answered fully.

Student A

(a) **(i)** A restriction endonuclease enzyme is selected ✓ which will cut the bacterial DNA either side of the *Bt* gene and certainly not within it. ✓ It is helpful if it leaves sticky ends. a

(ii) By using the same restriction endonuclease ✓ to cut the plasmid, leaving complementary sticky ends to which the bacterial DNA containing the gene is annealed using DNA ligase. ✓ a

(iii) Tomato cells have their cellulose walls enzymatically removed. ✓ *Agrobacterium tumefaciens*, containing the transgenic plasmid, will invade these cells, transferring its DNA into them. ✓ a

(iv) The modified plants will be resistant to insects ✓ and so fewer insecticides need to be used. ✓ a

(v) There are both moral and ethical objections. b

(b) (i) Normal mRNA reads UACGUAG, ✓ while anti-mRNA reads AUGCAUC. ✓ **c**

(ii) The RNA duplex contains the base uracil rather than thymine. ✓ DNA contains deoxyribose while RNA contains ribose. ✓ **c**

(iii) When the anti-RNA binds to the PG mRNA it is double-stranded, so it can't bind to the ribosome ✓ and there are no exposed bases for tRNA to bind to. ✓ **c**

(iv) If the tomatoes don't go soft quickly they are less likely to be damaged, so fewer will get wasted. ✓ **d**

e 15/18 marks awarded **a** All the marks are awarded. **b** This is not explanatory and fails to score. Student B has a good answer to this part of the question. **c** All correct. **d** While the student gets 1 mark for understanding that fewer tomatoes will be wasted, there is no link to profitability, so the second mark cannot be awarded.

Student B

(a) (i) Use a restriction enzyme ✓ to leave sticky ends. **a**

(ii) The plasmid would have been opened using a restriction enzyme ✗ and the bacterial gene joined using the enzyme DNA ligase. ✓ **b**

(iii) Using a gene gun. ✓ **c**

(iv) The plants will contain the toxin, which will kill any caterpillars feeding on them. ✓ **d**

(v) There is a danger that the gene for insect toxicity may be transferred to wild populations. ✓ The tomatoes may have an unforeseen effect on human health — for example, they may produce an allergic reaction. ✓ **e**

(b) (i) Normal sequence is UACGUAG. ✓ Antisense is TACGTAG. ✗ **f**

(ii) Both are double-stranded, but RNA contains uracil rather than thymine, ✓ and ribose in place of deoxyribose. ✓ **g**

(iii) The anti-mRNA inactivates the normal mRNA for PG translation. ✗ **h**

(iv) If the tomato doesn't go soft as quickly, it is less likely to be damaged ✓ when being transported and so fewer tomatoes will be thrown away. This makes the tomatoes more profitable. ✓ **i**

e 11/18 marks awarded **a** Leaving sticky ends is beneficial, but it is imperative that the enzyme cuts either side of the gene and not within it. 1 mark gained. **b** The precise answer is that the *same* restriction enzyme is used so that the exposed bases of the sticky ends are *complementary*. 1 mark awarded. **c** This is a correct method of inserting DNA into cells, but no detail is provided, e.g. that heavy metal pellets coated with recombinant DNA are fired through the surface of cells. 1 mark only. **d** This is only part of the answer — see Student A's response.

e These two suggestions are correct and well expressed for 2 marks. f The normal mRNA sequence is correct and earns 1 mark. However, the student gives the corresponding sequence of the antisense gene, not its mRNA (the anti-mRNA). g Both points correct, for 2 marks. h The student is just repeating information in the question and does not explain why no translation takes place. i This is worth both marks as it notes damage to tomatoes and links this to profitability.

Question 7 Haemophilia B: factor IX production and gene therapy

Haemophilia B is caused by a recessive allele of the *F9* gene leading to a deficiency of factor IX protein and prolonged blood clotting times.

(a) **Haemophiliacs are treated with factor IX obtained from transgenic sheep. The *F9* gene is cut from human DNA, annealed to a jellyfish gene and the combined DNA inserted into the nucleus of a sheep's egg cell. The egg is fertilised and the resultant modified embryo is implanted into a surrogate sheep. After birth, when the transgenic sheep lactates, its milk contains amounts of factor IX.**

(i) **The jellyfish gene, attached to the human *F9* gene, codes for a protein that glows green under fluorescent light. Explain the purpose of attaching this gene.** (2 marks)

(ii) **State a method of inserting the combined genes into a sheep's cell.** (1 mark)

(iii) **Many attempts to produce transgenic animals have failed. Of the live births, following implantation of embryos, very few are transgenic. Suggest one reason for this.** (2 marks)

(iv) **It is important that scientists still report the results from failed attempts to produce transgenic animals. Explain why.** (2 marks)

(b) **Trials of gene therapy for haemophilia B involve introducing the functional *F9* gene into liver cells where the gene is expressed.**

(i) **Gene therapy is more beneficial than treatment with injections of factor IX. Explain why.** (1 mark)

(ii) **State how the human *F9* gene would be introduced into liver cells.** (1 mark)

(iii) **While the results of various trials have been promising, there have been problems associated with gene transfer. Describe two problems.** (2 marks)

Total: 11 marks

e This question tests a range of assessment objectives. Notice that the command term *suggest* means that while you are not expected to know the answer, you should be able to use your biological understanding to provide an answer that is both relevant and plausible. The questions towards the end of each part involve an evaluation of the two treatments (AO3).

Student A

(a) (i) The jellyfish gene acts as a fluorescent marker ✓ identifying the egg cells that have taken up the *F9* gene. ✓

(ii) Genes can be inserted into animal cells by the process of electroporation. ✓

(iii) The transferred DNA may be damaged ✓ so that it is not expressed properly in the implanted embryo. ✓

(iv) So that the same procedures are not repeated ✓ and can be amended to provide a better approach. ✓ a

(b) (i) Gene therapy, if successful, is a permanent treatment so that injections do not need to be taken throughout the patient's life. ✓

(ii) An adeno-associated-virus (AAV) is used to introduce the *F9* gene. ✓

(iii) Some viral vectors, while altered to prevent disease, have become active and caused infection. ✓ Treatment may be short-lived and may need to be repeated at regular intervals. ✓ a

e **11/11 marks awarded** a Excellent answers throughout, for full marks.

Student B

(a) (i) The jellyfish gene glows under ultraviolet light. ✓ a

(ii) Using an electrical current to open pores in the cell membrane for DNA to enter. ✓ b

(iii) Very few implanted embryos survive. ✗ c

(iv) So that the same errors are not made ✓ which will save time and money in this research. ✓ d

(b) (i) Gene therapy is a once-and-for-all treatment whereas factor IX must be injected on a regular basis. ✓ e

(ii) Some trials used a retrovirus called lentivirus. ✓ f

(iii) Gene therapy often needs to be repeated. ✓ The viral-modified cells may be attacked by the immune system. ✓ g

e **8/11 marks awarded** a This is fine for 1 mark, but it does not explain its purpose in identifying those cells which have taken up the desired *F9* gene. b Correct for 1 mark. c This may be the case but the question refers to live births. See Student A's answer for an acceptable suggestion. d These two suggestions are correct and well expressed for 2 marks. e Correct. f Correct. g Both correct for 2 marks.

Question 8 Pharmacogenetics

Codeine is prescribed to patients for pain relief. However, it has to be activated in the body by enzymatic conversion to morphine before acting as a painkiller. The enzyme involved, CYP2D6, varies in its activity according to the genetic make-up of each patient.

The graph below shows the percentage of the UK population that have different activities of the CYP2D6 enzyme.

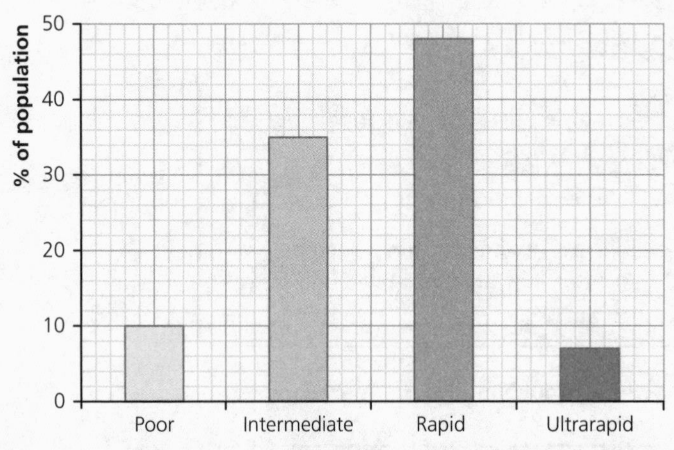

(a) **Explain the likely outcome for a patient who is a poor metaboliser treated with the normal adult dosage (40 mg).** (2 marks)

(b) **What percentage of the UK population would be liable to suffer from an overdosage of the painkiller? Explain your answer.** (2 marks)

(c) **Outline how patients needing pain relief could be genetically tested to ensure that the correct dosage of codeine is prescribed.** (3 marks)

(d) **The CYP2D6 enzyme is also responsible for the metabolism and elimination of 25% of prescribed drugs, for example metoprolol, a beta-blocker (given to reduce blood pressure).**

- **(i)** **Explain the effect of a rapid metabolising CYP2D6 enzyme in a patient prescribed metoprolol.** (2 marks)
- **(ii)** **Another drug, ritonavir, prescribed to deal with retroviruses, inhibits the action of CYP2D6. Explain the drug-to-drug interaction that would result if the two drugs were prescribed together.** (2 marks)

Total: 11 marks

e Much of this information will be unfamiliar to you. Study it carefully and make sure that your answers relate fully to the questions asked. Notice that the command term *outline* means that you are not expected to write about each stage in detail — just enough to make clear that you know the sequence of steps and can summarise each one.

Student A

(a) The patient may get little or no pain relief ✓ since morphine will be produced in small amounts. ✓ a

(b) About 7%. ✓ The ultrarapid metabolisers will act on codeine to produce high levels of morphine. ✓ a

(c) Genetically test the patient using appropriate gene probes in a microarray. ✓ This should allow the patient's alleles to be determined and whether they will produce poor or rapid metabolising enzyme. ✓ The dosage prescribed for the patient should be high for a poor metaboliser and low for a high metaboliser. ✓ a

(d) (i) The drug metoprolol will be metabolised too quickly. ✓

(ii) If ritonavir is taken along with metoprolol, then the metoprolol will not be broken down. ✓ There would be a danger of metoprolol having too great an effect, perhaps causing the blood pressure to drop dangerously low. ✓ a

e 10/11 marks awarded a Correct full answers throughout, except for (d) (i) where the answer is incomplete — Student B provides the full answer.

Student B

(a) The blood concentration of morphine does not reach high enough levels ✓ to be effective. ✓ a

(b) Approximately 7% are ultrarapid metabolisers. ✓ b

(c) DNA is extracted from the patient, made single-stranded and added to a microarray of probes which include those complementary to the different alleles of the gene producing the CYP2D6 enzyme. ✓ c

(d) (i) If the enzyme is a rapid metaboliser, it will break down metoprolol before it can act. ✓ The patient's blood pressure may not be reduced enough. ✓ d

(ii) The ritonavir will prevent the CYP2D6 enzyme from breaking down and eliminating the metoprolol. ✓ e

e 7/11 marks awarded a While both marks are awarded, the second point is only just given. It would have been better if Student B had said 'effective in providing pain relief'. b Correct for 1 mark, but no explanation for choosing ultrarapid metabolisers is given. c This is a more detailed description of the use of an appropriate microarray than required. However, the student does not outline the different steps in ensuring that the 'correct dosage of codeine is prescribed'. d Complete and well-expressed answers for 2 marks. e This is only part of the answer — see Student A's response.

Question 9 Haemochromatosis

Read the passage below and then use the information in the passage, and your own understanding, to answer the questions that follow.

Hereditary haemochromatosis, an inherited disorder of iron metabolism, is one of the most common genetic disorders in Ireland, affecting 1 in every 83 individuals. It is an autosomal recessive disorder. Individuals with haemochromatosis absorb excess iron. The result is iron overload and potential injury to organs, including the liver and heart. The symptoms vary among affected individuals, ranging from mild symptoms to life-threatening heart and liver disease.

The *HFE* gene, the gene responsible for hereditary haemochromatosis, was discovered in 1996 when chromosome 6 was analysed as part of the Human Genome Project. This gene encodes for the production of a transmembrane protein, called the HFE protein. The most common mutation of the *HFE* gene is called C282Y. The C282Y mutation replaces the amino acid cysteine with the amino acid tyrosine at position 282 in the HFE protein's amino-acid chain.

Studies of haemochromatosis have progressed through the use of knockout mouse models. Mouse models have been devised to investigate in which organ the gene acts. A mouse model engineered to lack the *HFE* gene in intestinal cells only shows no symptoms of haemochromatosis. However, a mouse model engineered to lack the *HFE* gene in the liver cells only shows all central features of the disease.

It seems that the HFE membrane protein detects the amount of iron in the body. In response to signals from the HFE protein, the liver cells produce the hormone hepcidin, which is released into the bloodstream and reduces iron uptake in the intestine.

(a) **Explain how hereditary haemochromatosis is normally inherited.** (1 mark)

(b) **Describe what is meant by the Human Genome Project.** (1 mark)

(c) **With respect to the *HFE* gene, describe what is meant by a 'knockout mouse'.** (1 mark)

(d) **Explain the C282Y mutation and the consequence of replacing cysteine with tyrosine in the HFE protein.** (2 marks)

(e) **Since haemochromatosis involves excessive absorption of iron, it used to be considered a disease of the intestine. However, it is now considered to be a disease of the liver. What is the evidence for this?** (2 marks)

(f) **Suggest what is happening in a person with haemochromatosis leading to iron overload.** (2 marks)

(g) **Hereditary haemochromatosis is not usually diagnosed until the age range 40–50 years by which time someone with the disease may already have had children. Suggest what advice a genetic counsellor could usefully offer to the family of someone diagnosed with hereditary haemochromatosis.** (2 marks)

(h) **Other hereditary disorders exist that have serious consequences in early childhood. In such cases, genetic testing for possession of the faulty allele is possible, either of cells from a fetus (pre-natal testing) or of cells from an adult (post-natal testing). While some people may emphasise the advantages of genetic testing, others will argue that such testing is not ethical. Discuss some of the issues involved.** (2 marks)

Total: 13 marks

ⓔ This is a fairly long passage and there is a lot of unfamiliar material to interpret. Read the passage carefully, underlining (or highlighting) new information. There is a variety of question types: part (b) requires a straightforward understanding of the Human Genome Project; parts (a) and (c) involve evaluation of information supplied in the passage; part (d) tests understanding of protein structure and is synoptic; parts (e) and (f) involve synthesis and the formulation of hypotheses; parts (g) and (h) involve discussion of social and ethical issues (AO3) — remember to give both sides of the argument.

Student A

(a) Both parents carry the haemochromatosis allele, each passing it on as a recessive allele. ✓ a

(b) The Human Genome Project determined the sequence of all the nucleotides on one set of chromosomes in the human cell. ✓ a

(c) A knockout mouse (for the *HFE* gene) has the *HFE* gene disabled in order to investigate the gene's role and the consequences for metabolism. ✓ a

(d) Replacing cysteine with tyrosine alters the tertiary structure of the protein ✓ since cysteine is involved in sulphur bridges. ✓ A change in the shape of the protein renders the protein non-functional. (✓) a

(e) A mouse model with the gene disabled only in intestinal cells shows no symptoms of haemochromatosis. ✓ A mouse model with the gene disabled only in the liver cells has the disease. ✓ a

(f) Too much iron is absorbed in the liver. ✗ a

(g) Children should be advised that they have the allele but would only be affected if the other parent was a carrier. ✓ They may be advised to see a doctor about genetic testing. ✓ a

(h) Embryos may be produced by IVF and tested so that only healthy embryos are implanted. ✓ This is called pre-implantation genetic diagnosis. a

ⓔ **10/13 marks awarded** a The answers provided are correct (with the maximum of 2 achieved in (d)), apart from (f), while (h) is incomplete. In (f), the student might have noted that 'lacking the HFE protein means that iron levels cannot be detected, so hepcidin is not produced (and so iron uptake is not controlled)'. Other relevant points in (g) and (h) are discussed in Student B's answer.

Student B

(a) On the Y chromosome. ✗ a

(b) This was the combined effort of many biologists to sequence all the nucleotides in the human genome. ✓ b

(c) The *HFE* gene has been made inoperative in a mouse for comparison with a mouse with a normal *HFE* gene. ✓ b

(d) The mutation occurred as a result of a mistake in DNA replication. ✓ c

(e) The disease is linked with the hormone hepcidin, which is produced in the liver. ✓ d

(f) There is an inability to produce hepcidin. ✓ e

(g) That the brothers and sisters of the person with haemochromatosis might also have the disease. ✓ f

(h) Genetic testing for the disease in a fetus would allow an affected fetus to be aborted. ✓ g

e 7/13 marks awarded a There is no evidence that it is confined to males. It is stated in the passage that the gene is located on chromosome 6. No score. b Both correct. c This is correct, for 1 mark. Other correct points are given in Student A's answer. d This is a relevant point, for 1 mark, but no reference is made to the findings from knockout mouse technology. Student A provides a fuller answer. e This is an appropriate suggestion and earns 1 mark. However, the student might have noted that this is due to the inability to detect iron levels (since the HFE protein is inoperative). f This is correct, for 1 mark. An additional piece of advice would be to provide a risk analysis, e.g. that sibling's chance of having the disease would be 1 in 4, assuming both parents to be heterozygous (the most likely event, especially if they failed to show symptoms of haemochromatosis). g A relevant point, but there are other issues — for example, some people do not approve of abortion, and the knowledge of being a carrier and that future pregnancies might result in abortion is extremely stressful.

Genes and patterns of inheritance

Question 10 Colour in budgerigars

In budgerigars, colour is determined by a gene with G and g alleles: the dominant allele G produces the wild-type green colour while the recessive g allele, in the homozygous condition, results in blue.

The expression of the G/g gene is influenced by the sex-linked *ino* gene (X^+/X^i). The presence of the dominant wild-type allele, X^+, allows the G and g alleles to be fully expressed (green or blue). However, the recessive X^i allele prevents their full expression: X^i (in the absence of X^+) along with the G allele results in yellow, while in combination with gg it produces albino.

Sex in budgerigars is determined by the X and Y chromosomes, though females are XY and males XX.

(a) **What term describes this type of gene interaction?** (1 mark)

(b) **Determine the genotypes of the following:**

- **green male**
- **blue female**
- **yellow male**
- **albino female**

(4 marks)

(c) **Two budgerigars of the following genotypes were crossed:**

$ggX^{+}Y \times GgX^{+}X^{i}$

Determine the expected proportions of offspring produced in terms of colour and gender. Show your working in a genetic diagram. (5 marks)

(d) **Explain how the genotype of a green female budgerigar ($GGX^{+}Y$ or $GgX^{+}Y$) may be determined.** (2 marks)

Total: 12 marks

ⓔ Genetics questions will usually use an unfamiliar context. In this case, while one genetic locus (colour) involves dominance, the other locus is sex-linked but the female is **XY** (as in all birds). In addition, there is gene interaction. This might seem tricky but as long as you have understood the information supplied and maintain your concentration, the question parts should be solved relatively easily. Part (b) helps you to collect your thoughts by asking for the genotypes of different phenotypes. In part (c), remember to show the *different* gamete types and then continue through the format of a genetic diagram to illustrate your answer.

Student A

(a) Epistasis ✓ a

(b) $GGX^{+}X^{+}$, $GgX^{+}X^{+}$, $GGX^{+}X^{i}$, $GgX^{+}X^{i}$ ✓

$ggX^{+}Y$ ✓

$GGX^{i}X^{i}$, $GgX^{i}X^{i}$ ✓

$ggX^{i}Y$ ✓ a

(c)

	GX^{+}	gX^{+}	GX^{i}	gX^{i}	Female gametes ✓ Male gametes ✓ Appropriate Punnett square ✓ Genotypes correct ✓ Phenotypes correct ✓ a
gX^{+}	$GgX^{+}X^{+}$ Green male	$ggX^{+}X^{+}$ Blue male	$GgX^{+}X^{i}$ Green male	$ggX^{+}X^{i}$ Blue male	
gY	$GgX^{+}Y$ Green female	$ggX^{+}Y$ Blue female	$GgX^{i}Y$ Yellow female	$ggX^{i}Y$ Albino female	

(d) Test cross with $ggX^{+}X^{+}$. ✓ If some offspring are blue then the parent is Gg whereas if they are all green then the parent is GG. ✓ a

ⓔ **12/12 marks awarded** a This excellent answer gains full marks.

Student B

(a) Epigenetic ✗ [a]

(b) Green male: GGX^+X^+ ✗

Blue female: ggX^+Y ✓

Yellow male: GGX^iX^i ✗

Albino female: ggX^iY ✓ [b]

(c)

	GX^+	gX^i	
gX^+	GgX^+X^+ Green male	ggX^+X^i Blue male	Female gametes ✓ Male gametes ✗ Appropriate Punnett square ✓ Genotypes correct for gametes ✓ Phenotypes correct for genotypes ✓ [c]
gY	GgX^+Y Green female	ggX^iY Albino female	

(d) This would require a test cross with the double recessive. ✓ [d]

e 7/12 marks awarded [a] This is epistasis — and illustrates the need to learn the terms used in biology. [b] Green males and yellow males have more than one possible genotype. The single genotype presented for each is not worth a mark. Always check that your answer is as complete as possible. [c] The male gametes (shown on the top row of the Punnett square) are incomplete — the law of independent assortment has not been applied — and so not worthy of a mark. However, this mistake is not penalised further. [d] This is not complete since the results of the test cross should have been worked out.

Question 11 Dihybrid inheritance: chi-squared test

An experiment to investigate the simultaneous inheritance of two characteristics was carried out using *Drosophila melanogaster* (fruit fly).

Female flies with sepia eyes and dumpy wings were crossed with male flies with red (wild-type) eyes and long (wild-type) wings. The F_1 all had red eyes and long wings. These were then interbred to produce an F_2. The results are shown in the table. A chi-squared (χ^2) test was used to determine the probability of the results departing significantly from the expected 9:3:3:1 ratio.

Phenotypes	Observed frequencies	Expected frequencies
Red eye, long wing	73	
Red eye, dumpy wing	19	
Sepia eye, long wing	31	
Sepia eye, dumpy wing	5	

(a) (i) **Complete the table to include the expected frequencies.** (1 mark)

(ii) **State the null hypothesis for the test.** (1 mark)

(iii) **State the degrees of freedom.** (1 mark)

(iv) **The χ^2 value was calculated as 4.223. State the probability value for the calculated χ^2 (using the statistics sheets).** (1 mark)

(v) **State your decision about the null hypothesis.** (1 mark)

(vi) **What can be concluded about the F_1 generation?** (1 mark)

(b) **State what can be deduced about the inheritance of sepia eye and dumpy wing with respect to the following genetic phenomena:**

- **dominance**
- **sex linkage** (2 marks)

(c) **State what can be deduced about the genotypes of the original parent *Drosophila*:**

- **sepia eyes and dumpy wings female**
- **wild-type (red) eyes and wild-type (long) wings male** (2 marks)

Total: 10 marks

e The results of a standard dihybrid cross are provided. Part (a) involves the use of the χ^2 test to determine the 'goodness-of-fit' to a 9:3:3:1 ratio. A2 Unit 2 papers may include statistical work and you should have practised the variety of tests repeatedly beforehand. Parts (b) and (c) require you to deduce genetic aspects of the inheritance of the two sets of alleles involved.

Student A

(a) **(i)** 72, 24, 24, 8 ✓

(ii) Any difference between the observed results and the expected results is due to chance. ✓

(iii) 3 ✓

(iv) $0.5 > p > 0.1$ ✓

(v) The null hypothesis is accepted. ✓

(vi) The F_1 flies are heterozygous at each of the two loci, e.g. +/se +/dp. ✓ a

(b) The sepia eye and dumpy wing alleles are both recessive. ✓ It is not possible to say if they are sex-linked. ✗ a

(c) The female parent is se/se dp/dp. ✓ The male parent is +/+ +/+. ✓ a

e **9/10 marks awarded** a All answers correct, apart from (b). The alleles cannot be sex-linked. If they had been, the F_1 flies would not all have been wild-type and a 9:3:3:1 ratio would not have been produced in the F_2 generation.

Student B

(a) **(i)** 72, 24, 24, 8 ✓ a

(ii) No significant difference in the results. ✗ b

(iii) 1 ✗ c

(iv) $0.05 > p > 0.01$ ✓ d

(v) The null hypothesis is rejected. ✓ e

(vi) They are heterozygous at the genetic loci. ✓ f

(b) Both sepia and dumpy are recessive. ✓ Neither is sex-linked. ✓ g

(c) The parents are heterozygous at one locus but homozygous at the other. ✗ h

e 6/10 marks awarded a Correct. b The null hypothesis for this test must specify that there is no significant difference between *the observed results and those expected* on the basis of theory. The student fails to score. c The correct answer is the number of classes (4) minus 1, i.e. 3. d This is correct for one degree of freedom (for which the student has already been penalised) and so is given the mark. e On the basis of the *p*-value given, this is correct and earns the mark. f This is correct. g Both answers are correct, for 2 marks. h This would not produce dihybrid F_1 flies. Student A has the correct answer; Student B fails to score.

Population genetics, evolution and speciation

Question 12 Polymorphism in ladybirds

Two-spot ladybirds, *Adalia bipunctata*, prey on aphids during the summer. They are normally red with two black spots, but a melanic form occurs which is black with four red spots. This polymorphism is shown in the diagram below.

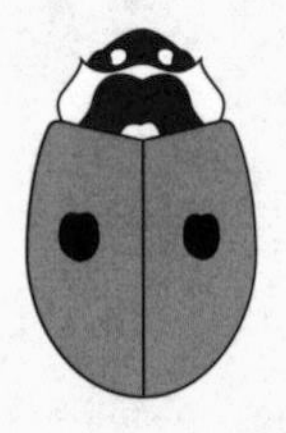

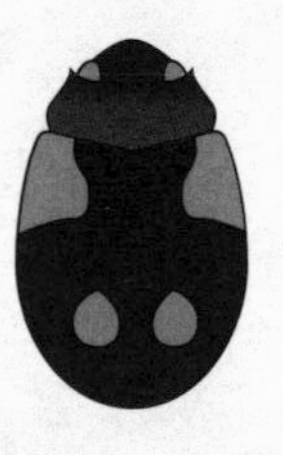

×5

The melanic form is more common in northerly regions where there are lower levels of sunshine. The black surfaces are more effective at absorbing radiant energy so that the body temperature rises rapidly and the ladybirds become more active.

(a) **Explain how natural selection might cause populations of ladybirds in northerly regions to possess relatively high frequencies of the melanic form.** (4 marks)

(b) **A population of the two-spot ladybird was investigated at Oxford Island. Out of a total of 500 sampled, 95 were found to be melanic. The melanic form is due to a dominant allele, B; the normal red form is homozygous recessive, bb.**

(i) **Calculate the proportion of the population that is represented by the normal red form.** (1 mark)

(ii) **Calculate the relative frequency of the typical (red–orange) allele and the melanic (black) allele, assuming the population to be in Hardy–Weinberg equilibrium. (Show your working.)** (2 marks)

(iii) **Calculate the relative frequencies of ladybirds that might be expected to be homozygous melanic and heterozygous melanic. (Show your working.)** (2 marks)

Two-spot ladybirds overwinter in sheltered positions where they remain inactive, living on stored fat. If activity is restored, their increased metabolism may cause their fat stores to be used up.

(c) The results of a study in Denmark, on the frequency of the b allele before and after the hibernating period, are shown in the graph below.

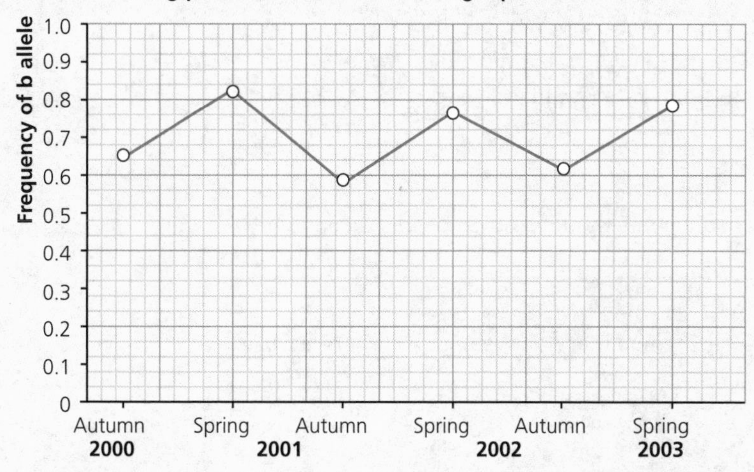

(i) The evidence from the graph shows that the melanic form was at a disadvantage and survived less well over winter. Explain this evidence. (2 marks)

(ii) Suggest an explanation for the poorer survival of melanic ladybirds over winter. (2 marks)

Total: 13 marks

ⓔ In part (a), read the information provided — melanic forms seem to have a selective advantage in northerly regions. You need to be able to order the steps by which this selective advantage will influence the allele frequency in the population; notice that the mark tariff is 4. In part (b), you are required to undertake a Hardy–Weinberg calculation; remember to start with the frequency of the homozygous recessive individuals. You are advised to 'show your working'. This means that if you make a slip in one phase of the calculation, you will not be penalised in subsequent phases, as long as you show correct arithmetical operations. Part (c) requires you to analyse and interpret the results of an investigation. Make use of the information supplied to help you construct an answer.

Student A

(a) The black forms are better at absorbing solar radiation and so are more able to catch prey, ✓ such as aphids. Thus females have more food for egg production, while males are more active in mating, i.e. they have greater fertility, ✓ and so more of their genes are transmitted to the next generation. ✓ The frequency of the melanic allele increases in the northerly populations. ✓ a

(b) **(i)** Frequency of red form = 405/500 = 81% = 0.81 ✓

(ii) $q^2 = 0.81$, so q (frequency of red allele) = 0.9 ✓
So p (frequency of black allele) = 0.1 ✓

(iii) Frequency of homozygous melanic = $p^2 = 0.01$ ✓
Frequency of heterozygous melanic = $2pq = 0.18$ ✓ a

(c) (i) The frequency of the b allele increased from autumn to spring in each of the years during the study, ✓ so the frequency of the B allele decreased and there were fewer black ladybirds in the spring. ✓

(ii) The black ladybirds were more likely to become active during the winter months. Their respiration rate would increase, ✓ depleting their fat reserves. ✓ a

13/13 marks awarded a Correct and well-thought-out answers throughout.

Student B

(a) In northerly regions the melanic form is more active and lays more eggs ✓ so is selected for. a

(b) (i) 95/500= 19% ✗ b

(ii) $q^2 = 81\%$, so $q = 9\% = 0.09$ ✗
So $p = 1 - 0.09 = 0.91$ ✓ c

(iii) Frequency of homozygotes = $p^2 = 0.91^2 = 0.83$ ✓
Frequency of heterozygotes = $2pq = 2 \times 0.91 \times 0.09 = 0.16$ ✓ d

(c) (i) There was an increase in the frequency of the red gene during each winter period. ✗ Therefore the black gene, and the number of black ladybirds, decreased. ✓ e

(ii) During mild periods during the winter black ladybirds become active and use up their food stores. ✓ f

6/13 marks awarded a This lacks sufficient detail and is worth only 1 mark. b The student calculates the frequency of the black phenotypes. The number of red ladybirds in the population is 500 less 95 — see Student A's answer. c The student works with percentages and obtains the wrong *q* value (the root of 81% or rather 0.81 is 0.9). However, this is not penalised further and 1 mark is awarded for understanding that $p + q = 1$. d Even though the values of *p* and *q* are wrong (already penalised), the student shows understanding of the frequencies of homozygotes and heterozygotes and obtains 2 marks. e Student B uses the term gene instead of allele, though this is penalised only once and so 1 mark is awarded. f 1 mark is awarded for suggesting depletion of food reserves, but the student should have noted that this would be due to an increase in respiration or metabolic rate.

Question 13 British and Irish stoats

Stoats and weasels are predators of rodents and rabbits. Their size can reflect their main prey. The weasel (*Mustela nivalis*), being the smallest, hunts mainly mice and voles, while the British stoat (*Mustela erminea stabilis*) hunts mostly rabbits, though the Irish stoat (*Mustela erminea hibernica*) will prey on a variety of rodents (shrews, mice and rats) as well as rabbits.

(a) Suggest why the British and Irish stoats are regarded as distinct subspecies while still being members of the same species. (2 marks)

To compare the size of the different species, 30 male members of each were captured and their length measured. From the data, the mean and standard deviation (error) of the mean was calculated for each species and the 95% confidence limits determined. The means and 95% confidence limits are shown in the graph below.

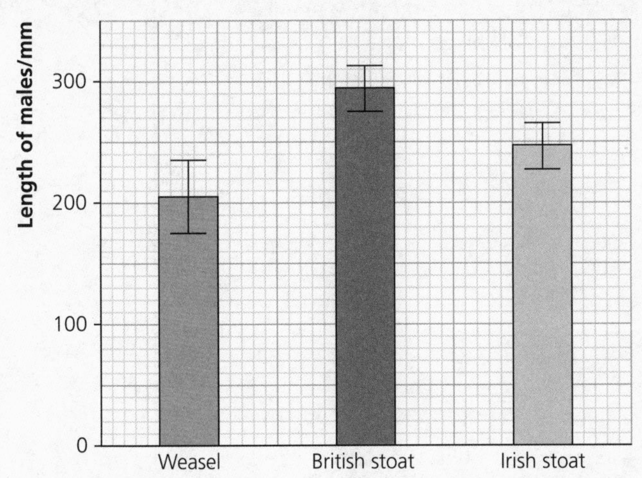

(b) Suggest why females were not included in the comparison. (1 mark)

(c) Compare the means and confidence limits for the three species. Do these suggest that there are significant differences in the lengths of the species? Explain your answer. (3 marks)

(d) A *t*-test was undertaken to compare the mean length of British and Irish stoats. The calculated *t* value was 3.644.

(i) State the null hypothesis for the test. (1 mark)

(ii) State the degrees of freedom. (1 mark)

(iii) State the probability value for the calculated *t* value (using statistics sheets). (1 mark)

(iv) State your decision about the null hypothesis. (1 mark)

At the end of the last ice age, stoats arrived into Britain and Ireland across land bridges from mainland Europe. The weasel arrived later into Britain, after Ireland had become separated from Britain and mainland Europe.

(e) Discuss how the lack of weasels in Ireland has affected the niches of the two stoat subspecies and their subsequent evolution. (3 marks)

Total: 13 marks

ⓔ This question involves an analysis and interpretation of the results of an investigation, though part (a) is about the concepts of species and subspecies. In parts (c) and (d), you are asked to complete statistical analyses: in (c) by comparing means and error bars and in (d) by interpreting the result of a *t*-test. You should have practised these skills well beforehand. Part (e) brings you back to evolutionary theory.

Student A

(a) They would be regarded as the same species because they are capable of interbreeding, ✓ but subspecies because there are sufficient distinctive differences between them. ✓

(b) Females might be nursing young and so it would be unfair to catch them. ✓

(c) The mean lengths for the weasel and the Irish stoat appear to be significantly different from the mean length of the British stoat, ✓ since the confidence limits do not overlap. ✓ However, there is uncertainty about whether the mean lengths of weasels and Irish stoats are significantly different since there is a degree of overlap in their confidence limits. ✓ **a**

(d) (i) That any difference between the means of the samples is simply due to chance. ✓

(ii) d.f. = 58 ✓

(iii) $p < 0.001$*** ✓

(iv) The difference is highly significant. ✓ **a**

(e) The original stoats arriving into Britain and Ireland were possibly the size of the Irish stoats. This intermediate size allowed them to take the wide range of prey species, including rabbits and rodents. ✓ In Britain, competition between the weasel and stoat ✓ caused them to evolve further, resulting in niche divergence, ✓ so that weasels specialised in smaller prey species with stoats hunting mostly rabbits. (✓) **a**

ⓔ **13/13 marks awarded** **a** These are excellent answers — full marks scored.

Student B

(a) If they could successfully interbreed they would be regarded as the same species. ✓ Neither would be able to breed with weasels. **a**

(b) Females could be a different size than males. ✓ **b**

(c) The means for the lengths of British and Irish stoats are more reliable than that for weasels. ✗ This is because their 95% confidence limits are narrower than the 95% confidence limits for weasels. ✗ **c**

(d) **(i)** There is no difference between the mean lengths of the two stoats. ✗ d

(ii) 58 ✓ e

(iii) $p < 0.001$ ✓ e

(iv) The null hypothesis is rejected. ✓ e

(e) Weasels and stoats in Britain have different niches — weasels hunt mostly small rodents and stoats chase mostly rabbits. ✓ The Irish stoat has a more general niche, taking a wide variety of prey. f

e 6/13 marks awarded a The first point is correct for 1 mark, while the second is irrelevant. In fact, there are distinctive colour and size differences between British and Irish stoats. b Correct for 1 mark. c The student does not understand '*significant*' difference. The role of confidence limits in determining reliability has been confused with their role in suggesting significant difference between sample means. d There is a difference. The null hypothesis is about there being no *significant* difference. e All these parts are correct. f Student B recognises the niche difference but fails to discuss how it may have evolved. 1 mark only.

Kingdom Plantae

Question 14 Ferns

(a) **State two differences between ferns and:**

- **mosses**
- **flowering plants** (4 marks)

(b) **The bracken fern (*Pteridium aquilinum*) grows in dense stands that rarely contain other plants. It is thought that phytotoxins leached by rain from the fern leaves are largely responsible for this suppression.**

(i) **The effects of bracken material on seed germination and seedling growth in three species of coniferous tree, Douglas fir (*Pseudotsuga menziesii*), Scots pine (*Pinus sylvestris*) and Sitka spruce (*Picea sitchensis*), were investigated. In field experiments, water-soluble extracts of bracken leaves were used to irrigate plots of seeds. In studies on seedling growth, other plots had bracken litter incorporated. In both cases, control plots had no bracken material added. The results are shown in the graphs overleaf.**

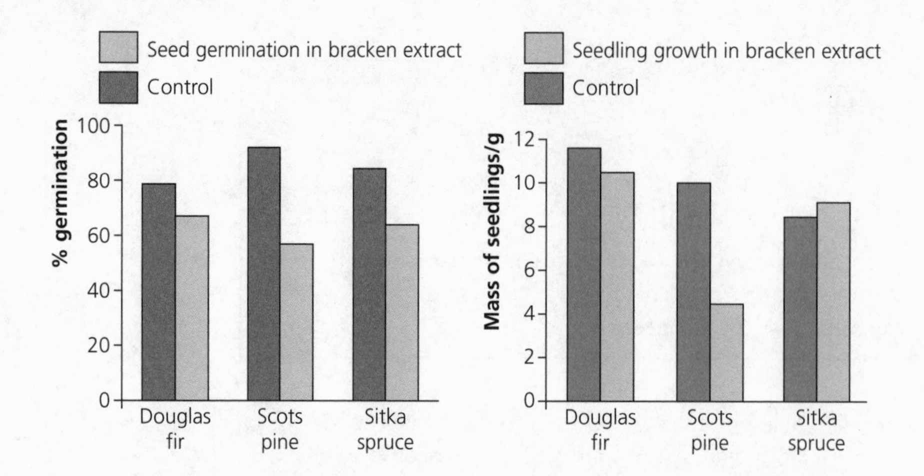

What conclusions can be made about the effect of bracken on the seed germination and seedling growth in the three species of coniferous tree? (4 marks)

(ii) Phytotoxins could be contained in any part of the bracken: the leaves, the stem or the roots. Design a laboratory experiment to test which of these parts, if any, exhibited the most inhibitory influence on seed germination. Use lettuce seeds as the test organism. (3 marks)

A number of chemical compounds, including cyanide, have been implicated in the various toxic properties of bracken to animals. Nevertheless, there are insects that eat bracken. These include larvae of sawfly species that feed on bracken leaves.

(iii) Wood ants have been shown to impose high rates of predation on larvae venturing onto the leaves. However, when sawfly larvae are attacked they respond by secreting a fluid that repels the ants. Suggest an explanation for the ability of sawfly larvae to repel foraging ants. (2 marks)

Total: 13 marks

e Part (a) requires straightforward understanding of differences between ferns and other plant divisions. Part (b) is more demanding. You are not expected to know about phytotoxins in bracken; this is unfamiliar information to test your ability to apply your skills in a novel situation. Read the information supplied carefully — twice if need be. In subpart (i), make sure you understand the graphs on seed germination and seedling growth; remember that small differences are unlikely to be significant. Subpart (ii) asks for a plan for an investigation. You need to address questions such as: what are the independent and dependent variables; what other variables do you need to control; what would your control experiment be; what replication is appropriate? Subpart (iii) tests your ability to synthesise a hypothesis.

Student A

(a) Ferns possess a cuticle, while the moss does not. ✗ Ferns have vascular tissue, while mosses do not. ✓ Ferns are dispersed by spores, flowering plants by seeds. ✓ Flowering plants have an upright stem, ferns do not. ✗ a

(b) (i) Bracken extracts inhibit seed germination, ✓ particularly of Scots pine, ✓ while the effect on Douglas fir is minimal. ✓ Seedling growth is reduced in Scots pine ✓ and to a lesser effect in Douglas fir. There is no inhibitory effect on seedling growth in Sitka spruce (✓) — if anything, bracken extract stimulates growth.

(ii) Leaves, stem and roots are removed from the fern and equal amounts ✓ are ground separately to make extracts. ✓ Lettuce seeds are planted in four different plots. One plot has no bracken solution added as it is used for comparison with the other plots, which are each treated with extract from a bracken part. ✓ Percentage germination is compared. (✓) Least germination indicates the bracken part that has the most inhibitory effect.

(iii) Sawfly larvae must be able to store the bracken toxins ✓ in such a way that the larvae's metabolism is unaffected. The toxin in the fluid released by the larvae when they are attacked is detected by the ants, ✓ which are repelled. b

e 11/13 marks awarded a The first statement is incorrect since mosses do have a cuticle but it is confined to the spore-producing capsule. The second and third statements are correct for 2 marks. The final statement is incomplete since the student fails to say that ferns do have a stem but that it is a rhizome. b Student A scores maximum marks in part (b).

Student B

(a) Ferns have true roots, while mosses have rhizoids for anchorage. ✓ Ferns have leaves with a cuticle, while in mosses the cuticle is confined to the capsule. ✓ Ferns have a horizontal stem, a rhizome, while in flowering plants the stem is upright. ✓ Ferns reproduce by spores, while flowering plants reproduce by seeds. ✓ a

(b) (i) Bracken extracts have a negative effect on seed germination. ✓ Seedling growth is negatively affected in Scots pine. ✓ b

(ii) Remove leaves, stems and roots from a bracken plant and use a mortar and pestle to grind up the parts separately. ✓ Add the extract to plots of lettuce seeds. Compare the percentage germination of the seeds treated with extracts from different bracken parts. ✓ c

(iii) The ants are afraid of the fluid secreted by the larvae. ✗ d

e 8/13 marks awarded a Correct points for 4 marks. b There are two correct responses, though the student does not make any conclusions about the relative effects on the different tree species. c There is no control of variables (e.g. equal mass of plant parts) and no control experiment (lettuce seeds with just water added, i.e. lacking extracts). 2 marks are scored. **d** This answer lacks any detail and the student fails to score. Student A provides a correct answer.

Kingdom Animalia

Question 15 Flatworms and earthworms

The New Zealand flatworm (*Arthurdendyus triangulatus*) is an invasive earthworm predator that, having first been found in Belfast in 1963, has now colonised much of Northern Ireland. Three factors have influenced its spread: the presence of earthworms that make up the bulk of its diet, a humid or damp microclimate, and a soil temperature that does not exceed certain values.

(a) **Flatworms are members of the phylum Platyhelminthes, while earthworms are members of the phylum Annelida.**

(i) **State one feature which flatworms and earthworms have in common.** (1 mark)

(ii) **State three features which differ between flatworms and earthworms.** (3 marks)

(b) **Flatworms and earthworms both inhabit damp environments. Explain why.** (1 mark)

(c) **The effect of exposure to a wide range of temperatures on the survival of the New Zealand flatworm was investigated. The results are shown as mortality curves in the graph below.**

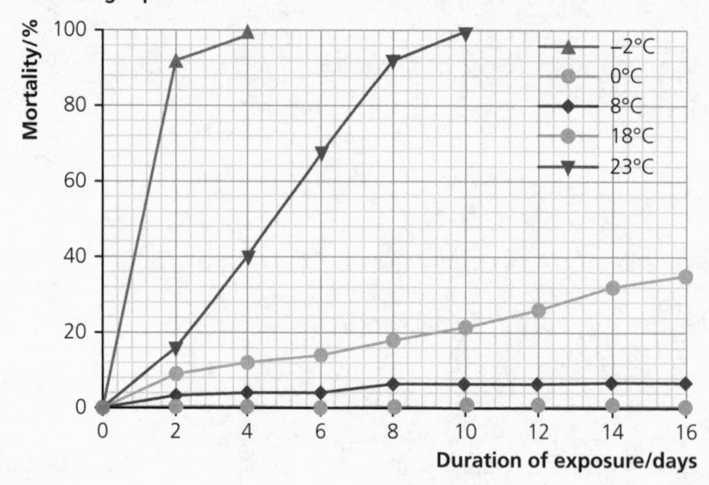

What conclusion can be made from these results? Suggest how the results explain the spread of the New Zealand flatworm through Northern Ireland. (4 marks)

Total: 9 marks

e Part (a) is relatively straightforward, asking you to compare and contrast platyhelminths and annelids. Part (b) is simple though synoptic in nature. Part (c) is more demanding, requiring an analysis and evaluation of the results of an investigation.

Student A

(a) (i) Both are bilaterally symmetrical. ✓

(ii) Platyhelminths have a sac-like gut, while annelids have a through-gut. ✓ Annelids are metamerically segmented, while platyhelminths are not. ✓ Annelids have a blood system, while platyhelminths do not. ✓ a

(b) Their body surface is moist to allow gas exchange. ✓ a

(c) Extreme temperatures increase the mortality of the flatworms. ✓ In fact, the lowest and highest temperatures cause 100% mortality within 10 days. ✓ Temperatures in the range 8–18°C have low mortality rates during the experimental period, ✓ though prolonged exposure at all temperatures leads to increased mortality. ✓ Northern Ireland has only short periods of extreme temperature throughout the year and is relatively easy for flatworms to colonise. (✓) a

e **9/9 marks awarded** a Excellent, detailed answers throughout.

Student B

(a) (i) They have a moist, permeable surface. ✓ a

(ii) Flatworms have a blind-ending gut and earthworms have a one-way gut. ✓ In flatworms the gut is simple though branched and in earthworms the gut is regionally specialised. ✓ b

(b) Because they have a moist surface. ✗ c

(c) Temperatures of 0°C and 23°C kill flatworms. ✓ Temperatures of 8°C and 18°C have little effect on the mortality of flatworms. ✓ Northern Ireland has a similar climate to New Zealand. d

e **5/9 marks awarded** a Correct for 1 mark. b There are two correct responses, though these relate only to the gut. Student A provides other correct responses. c This is not a sufficient explanation — see Student A's answer. d This answer lacks detail and only two points are worthy of marks.

Section B Essay questions

Question 16 PCR and genetic fingerprinting

(a) Outline the technique of the polymerase chain reaction (PCR). (6 marks)

(b) Write an essay on genetic fingerprinting (DNA profiling). (12 marks)

In this question, you will be assessed on your written communication skills.

Total: 18 marks

ⓔ Notice that part (a) asks you to *outline*, meaning that you should restrict yourself to the main points, while part (b) asks you to *write an essay*, meaning that you should be more expansive. In both parts you should take some time to *devise a plan*. The mark scheme for each part is divided into three bands. An answer that demonstrates limited content, which lacks structure and fails to use correct biological terminology will be placed in the lower band. An answer that is articulate and integrated, which correctly links the relevant points and uses correct biological terminology, will be placed in the upper band. Answers which fall in between these two descriptions are placed in the middle band.

Student A

(a) The polymerase chain reaction, PCR, is used to amplify minute quantities of DNA. ✓ The technique involves mixing the DNA with various reagents in a tube placed in a machine called a thermal cycler. ✓ The mixture includes primers, free DNA nucleotides and thermostable DNA polymerase. ✓ In the machine, the temperature is raised to 95°C to separate the two strands of the DNA. ✓ The mixture is cooled to 50°C to allow the primers to bind to the ends of the DNA strands. ✓ This prevents the DNA strands rejoining and acts as a starting point for the polymerase enzymes. ✓ The mixture is heated to 72°C and the thermostable polymerase copies each strand, starting at the primers. ✓ The process is repeated, and with each cycle the quantity of DNA is doubled. **a**

(b) In the genome, there are highly variable regions where the nucleotide sequences are repeated many times. These repeat sequences are found in the non-protein-coding DNA and are called satellite sequences. ✓ Microsatellites contain 2–5 base pairs repeated between 5 and 15 times, and are highly distinctive in any individual. ✓ It is these regions which are analysed to produce a 'fingerprint' which is unique for each person.

The traditional method of genetic fingerprinting cuts satellite repeat sequences from the DNA using restriction endonucleases. ✓ These must be selected to cut the DNA at sites either side of the satellite region. ✓ If a satellite region has many repeat sequences then it will be long. ✓ A short satellite region means that there are few repeats. The fragments of DNA, some of which will contain satellite regions, are then separated using gel electrophoresis. ✓ In an electrical field in a gel, smaller fragments will move further with longer fragments left behind. ✓ The separated fragments are heat-treated to make them single-stranded and then transferred to a nylon sheet by a technique called Southern blotting. ✓ The DNA fragments are not visible at this stage, but only those with the satellite regions need to be identified. This is done using fluorescently labelled DNA probes ✓ which have been synthesised so that they will hybridise with those fragments containing the selected complementary repeat sequences. ✓ The satellite fragments with fluorescent probes attached will show up under ultraviolet light giving a pattern like a bar code unique to each individual. ✓

A newer technique, referred to as DNA profiling, involves placing fluorescently labelled probes either side of selected microsatellites ✓ and copying these using the PCR. ✓ The copied fragments are separated using gel electrophoresis and, since they have a fluorescent tag, will show up using a laser scanner. ✓

Genetic fingerprinting, or DNA profiling, is a powerful tool in forensic science, ✓ in settling paternity disputes and in establishing the evolutionary relationship between different groups of organisms. ✓ a

ⓔ **18/18 marks awarded** a Student A has scored a high number of indicative points in the mark schemes for both parts and is placed in the top band. Since ideas are clearly and sequentially presented and good use has been made of appropriate terminology, a maximum of 18 marks is awarded.

Student B

(a) The DNA extracted from a trace of blood at a crime scene can be minute. The polymerase chain reaction increases the quantity of this DNA ✓ so that there is sufficient for further analysis, for example to produce a genetic fingerprint. Basically the DNA is heated to separate its strands ✓ and a heat-stable polymerase enzyme added ✓ which replicates each strand. The process is repeated and the number of DNA molecules is doubled with each repetition. a

(b) Like fingerprints, DNA is unique, with the exception of identical twins. No two people have the same DNA. Genetic, or DNA, fingerprinting is a technique that can identify the person responsible for a violent crime from the physical traces left at the scene. ✓ The DNA can be extracted from samples of blood, hair, skin or semen. The sample DNA is cut into fragments at multiple sites with restriction endonucleases. ✓ These fragments of DNA are then separated using gel electrophoresis. ✓ The size of fragments produced varies from individual to individual. The fragments are tagged using radioactive or fluorescent probes. ✓ These tagged fragments are then viewed using X-ray film if the probes are radioactive, or using ultraviolet light if using fluorescent probes. ✓

Genetic fingerprinting has proved valuable, not only in convicting criminals but also for establishing maternity or paternity and proving family relationships. ✓ a

ⓔ **10/18 marks awarded** a Part (a) lacks the necessary detail about the PCR — for example, the involvement of primers and the variation in temperature are not specified. Again, in part (b) important detail is missing — for example, there is no mention of the variability in non-coding DNA, microsatellites or repeat sequences. However, there are sufficient indicative points for Student B to be placed in the middle band. Ideas are well sequenced and a reasonable number of appropriate scientific terms are used so that a final mark of 10 is awarded.

Question 17 Evolution and speciation

Write an essay on how evolutionary change may take place in populations and how this could lead to speciation. (18 marks)

In this question, you will be assessed on your written communication skills.

Total: 18 marks

ⓔ There are two aspects noted in the title of this essay — what is involved in evolutionary change and what is further involved for speciation to take place. You should also discuss the situation in which evolutionary change and/or speciation may not occur. You should spend a short time organising a plan for your essay and ensure that you are going to provide the necessary detail, while keeping your comments relevant. The essays are marked according to the number of indicative points in the content and placed in one of three bands; the final mark is then based on the quality of written communication. The quality of written communication is assessed on the clarity and coherence of your writing and on the use of appropriate specialist vocabulary.

Student A

If the members of a population were the same, evolution would not be possible. However, several processes occur to ensure that populations are genetically variable. ✓ Natural selection acts on this variation. ✓ For example, a population of plants may be adapted differently to the water content of the soil. If conditions are in general moist and the environment stays the same, then plants adapted to moist conditions will be favoured. However, if the environment changes and conditions become drier ✓ then those variants adapted to drier conditions — and not previously favoured — would be selected for. ✓ This is called directional selection. ✓ This means that they will be more likely to survive ✓ and reproduce, ✓ so the frequency of 'dry' alleles will increase in the population. ✓ This change in allele frequency represents evolution. ✓ The end result is that the population stays adapted to its environment. ✓

In the theory of allopatric speciation populations become geographically isolated. ✓ For example, some plants may, somehow, have been moved to an island. The island may be wetter: that is, the environmental conditions differ. ✓ This means that selection pressures in the original and new environments differ. ✓ The result is a divergence of the gene pools of the two populations. ✓ If this takes place over a long enough time, sufficient genetic differences may occur ✓ to make the populations reproductively incompatible. ✓ If the two populations are unable to breed ✓ they are known as different species ✓ and speciation has occurred. **a**

Quality of written communication. **b**

ⓔ **18/18 marks awarded** **a** The high number of indicative points puts the student in the top band, **b** and the coherence of the essay, including the use of correct biological terminology, means that a maximum of 18 marks is awarded.

Student B

Natural selection acts on the genetic variation of a population. ✓ Genetic variation results from meiosis. This is because of crossing over and the independent assortment of chromosomes. Crossing over happens when chiasmata form during prophase I of meiosis. Independent assortment happens because of the random alignment of homologous chromosomes during metaphase I. Both stabilising selection ✗ and directional selection cause evolutionary change. ✓ The type that is selected survives and is favoured. ✓ If the environment changes, a type, not previously favoured, may be selected. ✓ This type will reproduce and pass on more genes to the next generation ✓ so there will be an increase in allele frequency. ✓

Speciation occurs when populations become geographically separated. ✓ Two species are formed when they can no longer interbreed. ✓ a

Quality of written communication. b

e **9/18 marks awarded** a Too much of this answer is not relevant, i.e. the detail on genetic variation. There is confusion about the effect of stabilising selection and, in general, there is insufficient detail, particularly with respect to speciation. Nevertheless, there are sufficient appropriate points for the student to be placed in the middle band. b While not always relevant, material is organised with reasonable coherence and there is some use of appropriate specialist vocabulary. A final mark of 9 is awarded.

Knowledge check answers

1 It contains ribose, so it is derived from a nucleotide of RNA.

2 Four ATPs are produced per glucose: glucose yields two glycerate bisphosphates, each of which produces two ATPs in its conversion to pyruvate. However, two ATPs are used in the first reaction of glycolysis so the *net* production per molecule of glucose is two molecules of ATP.

3 Triose phosphate is oxidised (hydrogen removed); NAD is reduced to NADH.

4 NADH, ATP, pyruvate

5 **(a)** Pyruvate dehydrogenase removes hydrogen from pyruvate.
(b) Pyruvate decarboxylase removes CO_2 from pyruvate.

6 **(a)** 2
(b) 6

7 CO_2 diffuses out of the cell, into the blood and then into the alveoli, eventually to be expired.

8 **(a)** Cytoplasm
(b) Mitochondrial matrix
(c) Inner mitochondrial membranes (cristae)

9 **(a)** Oxygen is the final electron acceptor; if not available, all carriers are reduced.
(b) If the hydrogen carriers, NAD and FAD, are reduced then they are not available to act as coenzymes in the dehydrogenase reactions of the Krebs cycle.

10 In chemiosmosis: 2.5 ATPs per NADH; 1.5 ATPs per FADH

11 Folding increases the area of the membrane, so increasing the number of membrane-bound electron carriers.

12 73.1 kg

13 Glycolysis metabolises glucose to pyruvate; anaerobic respiration has *additional* reactions to form either lactate (in animals) or ethanol and CO_2 (in plants and fungi).

14 Lactate and ATP

15 The acidic pH has an adverse effect on enzymes (specifically the ability of substrates to bind to the active sites).

16 This is because the CO_2 has diffused out of the cell and because plant and yeast cells lack the specific enzyme to reattach it to ethanol.

17 0.69

18 Mitochondria and nuclei

19 **(a)** Green (wavelength of about 550 nanometres)
(b) Black

20 Different pigments absorb at slightly different wavelengths, thus increasing the range of wavelengths absorbed.

21 **(a)** Used to reduce NADP

(b) Transferred through the electron transfer chain (and replace those 'lost' from photosystem I)

(c) Replace electrons 'lost' from photosystem II

22 Similarities: both derive energy from the transfer of electrons along an electron transport chain; both result in the synthesis of ATP.

Differences (*any two from*): oxidative phosphorylation occurs in mitochondria whereas photophosphorylation occurs in chloroplasts; oxidative phosphorylation has a reduced coenzyme (e.g. NADH) as an electron source whereas photophosphorylation has light-activated pigments as an electron source; the final electron acceptor in oxidative phosphorylation is O_2 whereas the final electron acceptor in photophosphorylation is photosystem I (and then NADP).

23 Ribulose bisphosphate, ribulose bisphosphate carboxylase (rubisco), glycerate phosphate

24 Reduced light intensity means less ATP and NADPH for the conversion of glycerate to triose phosphate. Since this is a vital step in the light-independent stage, its rate is reduced.

25 Light

26 There is a higher concentration of CO_2.

27 It produces CO_2, which is fixed in the light-independent stage of photosynthesis, and increases the temperature, thereby increasing the rate of enzyme-controlled reactions.

28 A triplet code produces sufficient codes; a doublet produces only 16 codes, which is insufficient for 20 amino acids.

29 146

30 Similarities (*any two from*): both involve breaking hydrogen bonds to separate the DNA strands (involve helicase); both involve a DNA strand acting as template; both occur in the nucleus.

Differences (*any two from*): transcription involves only one DNA strand as template whereas DNA replication involves both strands; transcription involves RNA polymerase whereas replication involves DNA polymerase; transcription involves ribonucleotides whereas replication involves deoxyribonucleotides.

31 Six polypeptides, produced by E1E2E3, E1E3E2, E2E1E3, E2E3E1, E3E1E2 and E3E2E1.

32 Thr (threonine)

33 **(a)** GUCUGUAAG

(b) CAG, ACA, UUC

(c) Val, Cys, Lys

34 **(a)** Transfer RNA

(b) Ribosome

(c) Peptide

35 If it is the third base which is substituted this may not affect the amino acid being coded for, e.g. Ala (alanine) is coded for by GC*x* (see Table 3).

36 A base substitution affects only one triplet code; all the other codes remain the same.

37 Differences are due to epigenetic changes accumulating over time.

38 Oncogene, since methylation silences genes and so a decrease in methylation would allow a gene to be activated.

39 Because the replication takes place at 72°C.

40 DNA helicase

41 GATA repeated four times

42 They are unique to each person and therefore allow individuals to be distinguished.

43 Both are short, single-stranded sequences of DNA, complementary to the DNA strand to which they will attach. However, primers are used in the PCR to attach to the ends of DNA strands, allowing binding of DNA polymerase since it cannot bind directly to single-stranded DNA, while probes are used to detect the presence of a base sequence in a strand of sought-after DNA, and labelled to allow their location to be identified.

44 Because they may have only a minute sample to begin with.

45 The nucleotide (base) sequence in a haploid set of chromosomes and both sex chromosomes (if present).

46 CTCGATGA (or reverse, depending on which is the 3′ end).

47 Different alleles have different base sequences and so code for different amino acid sequences, leading to different tertiary structures and enzyme activities if the structural difference affects the active site.

48 A lower dose, since codeine will be rapidly activated and the patient will be vulnerable to overdosing.

49 A restriction enzyme has a specific active site, complementary to a particular sequence of 4–8 nucleotides (bases). Therefore, different restriction enzymes cut DNA in different specific places.

50 Because a single restriction enzyme may cut at a base sequence in a gene, or cut either side of several genes.

51 **(a)** One strand will contain the sequence TATCGTCAGC.

(b) The probe will be either fluorescent or radioactive.

52 There are many copies of mRNA in the cell and there is no need to remove introns.

53 The transformed bacteria would be resistant to kanamycin but not to chloramphenicol.

54 A gene for antibiotic or herbicide resistance — only the transformed cells will survive; or a fluorescent marker gene — transformed cells will glow under ultraviolet light.

55 **(a)** Restriction enzyme
(b) Gene (DNA) probe
(c) DNA ligase
(d) Reverse transcriptase (and DNA polymerase)

56 The defective allele will always be expressed.

57 Because the lipid layers of liposome and membrane readily fuse (just as secretory vesicles fuse naturally with the cell-surface membrane).

58 Pollen grains do not possess chloroplasts.

59 Microorganisms are used that are unable to grow outside the controlled environment of the laboratory (using strains with low growth rates or with 'suicide' genes); use of efficient filters to prevent the escape of microbes into the atmosphere.

60 Heterozygous

61 Because the parents are heterozygous; test cross with a curled-winged fruit fly.

62 Incomplete dominance

63 **(a)** Since the heterozygote is lethal, the allele is rapidly eliminated from the population; the allele reappears only through recurrent mutation.
(b) Huntington's disease is late-onset, so individuals carrying the allele can pass it on to their offspring.

64 1 hairless:1 normal

65 Baby with blood group O belongs to parents Y; baby 'B' to parents X; baby 'AB' to parents W; baby 'A' to parents Z.

66 Yes, if the mother was a carrier and the father haemophiliac; X^hX^h.

67 GgX^BX^b, green-bodied, black-striped male; ggX^BX^b, blue-bodied, black-striped male; GgX^bY, green-bodied, brown-striped female; ggX^bY, blue-bodied, brown-striped female.

68 Eight: ABC, ABc, AbC, Abc, aBC, aBc, abC, abc.

69 AaEe (agouti), Aaee (white), aaEe (black) and aaee (white); the ratio is 1 agouti:2 white:1 black.

70 512

71 **(a)** Meiosis, fusion of gametes (especially cross-fertilisation) and mutation
(b) Reduces the range

72 **(a)** Frequency of **C** allele increased
(b) Rate of mutation not affected

73 Directional selection

74 Populations on an island are geographically isolated; different selection pressures cause genetic divergence of the island and mainland populations; divergence may lead to the populations being reproductively isolated — when this happens the populations have become different species.

75 Because pollutants have direct access to all tissues since there is no cuticle.

76 Humid (moist) environment; tolerant of dim light conditions.

77 Have upright stems so that leaves (and flowers) are carried higher into the canopy, especially with the evolution of trees; some species have xerophytic adaptations, allowing them to live in dry habitats.

78 No cell walls; possess a gut (enteron) in which digestion takes place; have a nervous system and can move about.

79 Platyhelminths have bilateral symmetry, while cnidarians are radially symmetrical.

80 Radial symmetry is the arrangement of the body around a central axis; bilateral symmetry is the arrangement of the body into right and left halves that are mirror images of each other.

81 They have a moist surface (for gaseous exchange) and lack a skeleton (against which locomotory muscles might act).

82 Cylindrical shape provides a high surface-area-to-volume ratio and a relatively low level of activity means they have low rates of oxygen consumption.

83 Animals which are sessile show radial symmetry, allowing them to respond to the environment (e.g. during feeding) in all directions. Bilateral symmetry is associated with animals which move through their environment in search for food.

Index